THE TRUTH ABOUT THE HEAVENS

Michael Cyrek

VANTAGE PRESS
New York / Atlanta
Los Angeles / Chicago

FIRST EDITION

Published by Vantage Press, Inc.
516 West 34th Street, New York, New York 10001

Manufactured in the United States of America
ISBN: 0-533-07236-0

Library of Congress Catalog Card No.: 86-90300

Contents

Illustrations v

Introduction vii

Chapter 1. **Sunlight and the Photon** 1
Chapter 2. **Sunspots and Related Objects** 10
Chapter 3. **The Aurora Borealis and the Van Allen Belts** 34
Chapter 4. **The Perpetual Universe** 44
Chapter 5. **The Big Bang** 63
Chapter 6. **Black Holes** 84
Chapter 7. **The Wave Nature of Particles of Matter** 91

Addendum 109

Bibliography 115

Index 117

Illustrations

1. Experiment with Light 9
2. Sunspot Activity Cycle 12
3. Cause of Sunspot Cycle 13
4. Ring Nebula in the Constellation Lyra (NGC 6720, M57) 15
4a. Possible Type of Explosion for Ring Nebula and NGC 7293 16
5. The Helix Nebula in Aquarius (NGC 7293) 17
6. Owl Nebula in Ursa Major (NGC 3587, M97) 19
6a. Possible Type of Explosion for Owl Nebula and NGC 2392 20
7. NGC 2392 Planetary Nebula in Gemini 21
8. Dumbbell Nebula in Vulpecula (NGC 6853, M27) 22
8a. Possible Type of Explosion for Dumbbell Nebula 23
9. The Crab Nebula in Taurus (NGC 1952, M1) 25
9a. Possible Type of Explosion for Crab Nebula 26
10. Cepheid Variable 28
10a. Wave Pattern of Cepheid Variable 29
11. Cygnus A Radio Source 30
11a. Cygnus A Radio Lobes 31
11b. Cygnus A as Ring Nebula 31
12. Galaxy M87 (NGC 4486) in Virgo, Virgo A Radio Source 33
13. Aurora Borealis Dual Bands 37
14. Aurora Borealis Single Band 38
15. Aurora Borealis Coronal Rayed Band 39
15a. Aurora Borealis Coronal Rayed Band 40
16. Aurora Borealis Arc 41
17. Van Allen Radiation Belts 42
18. Classification of Galaxies 51
19. Giant Elliptical Galaxy in Virgo (NGC 4486, M87) 54
20. Galaxy, Type E0 (pec) in Centaurus, NGC 5128 56
21. Galaxy, Type Sa/Sb in Virgo (NGC 4594, M104) 57
22. Galaxy Type Sb in Coma Berenices, NGC 4565 59

23. Galaxy Type Sc in Triangulum (NGC 598, M33) 60
24. Galaxy, Barred Spiral in Pegasus, NGC 7741 62
25. Basic Principle of Red Shift 64
26. Visible Light Spectrum 65
27. Galaxy, NGC 7603 67
28. Galaxy, NGC 1199 68
29. Great Galaxy in Andromeda (NGC 224, M31) 70
30. Probable Orbits of Andromeda and Milky Way 71
31. Comparison of Temperature Scales 73
32. Velocity Curve of Matter 76
33. Big Bang Universe during First 100 Years 78
34. Big Bang Universe after First 300 Million Years 78
35. Determining the Age of Big Bang Universe 80
36. Dimensions of Big Bang Universe at Present Time 81
37. Present Shape of Visible Universe 82
38. Black Hole 85
39. Possible Visibility of Black Hole 86
40. Distribution of Heat within a Star 88
41. Magnetic Field around Wire, Angular View 94
41a. Magnetic Field around Wire, In-line View 94
42. Magnetic Field around Bar Magnet 95
42a. Magnetic Field of Horseshoe Magnet 95
43. Principle of Electric Motor 96
44, 44a, 44b. Oscillating Photons in Space 97, 98
45. Wave Created by Photon 98
46. Simple Oscillating Circuit 99
47, 47a, 47b. Wave Change of Particle beyond Velocity of Balance 102, 103
48. Nature of Wave Change of Particle 104
49. Spinning Protons, Unstable Axis Position 105
49a. Spinning Protons, Stable Axis Position 106
50. Magnetic Fields around Spinning Neutron 107
51. Magnetic Fields around Neutron in Flight 107
52. Circular and Elliptical Orbits 110
53. Parabola 111
54. Hyperbola 112

Introduction

This book is written to try to enhance the interest of the average reader and the general public in astronomy and some other sciences. The book is written for the understanding of the average reader. The public will find astronomy and the other sciences fascinating subjects to read about and, in so doing, will broaden its intellectual field of knowledge.

This book will also be of great interest to the professional scientist because of the new ideas it introduces. These new concepts should open new avenues of thought and new directions for research. It is not meant to offend the professional scientist because of its refutation of some current scientific opinions, but I believe more realism should be introduced into theoretical thinking.

Astronomy is the main topic of discussion with one chapter devoted to physics and another to the earth's atmosphere. It is devoted only to selected subjects that I believe need clarification. For a complete coverage of astronomy, I suggest the reader get a regular astronomy book so that a better understanding can be had.

One could also get other books for reference when reading the chapter on the aurora borealis and the Van Allen radiation belts.

On the wave nature of particles, one would need a physics book.

Do not rush through this book, because it may need a reasonable amount of study to understand. Although I've written this book in the simplest possible terms, the subjects may be a little difficult for the average reader to understand. However, it can be understood.

The most interesting chapters are 4 and 5. Chapter 4 explains how the galaxies renew themselves, and in so doing, we have a perpetual universe. Chapter 5 explains the illogical concept of the big bang universe and why we should abandon that line of thinking. Chapter 7 is an interesting chapter on the wave nature of particles. Chapter 1 may be the most controversial in determining how the sun's energy is generated, but I'm inclined to think there is some truth in it even though it reintroduces an old idea: that gravity is a source of the sun's energy. Chapter 2 also reintroduces an old idea that was supposedly disproved. However, my explanation is expanded to prove its worthiness.

The rest of the book contains a lot of new information not found in any other astronomy book.

Chapter 1

Sunlight and the Photon

What causes the sun to radiate so much light? Contemporary science says that atomic energy is the cause. Although atomic energy may contribute some energy deep in the central core of the sun, I believe that the enormous size of the sun and its resultant gravitational force is the greatest reason why the sun radiates light.

This gravitational force is so enormous that it actually liquifies the hydrogen gas that surrounds it. This gas constitutes the very thick outer layer of the sun. There are smaller quantities of other gases that form successive layers deep down below the hydrogen layer, but hydrogen is by far the most predominant component of the sun. Then, below these other gases, light solids start forming their layers.

The light solids are compressed to enormous densities. As we go further toward the center of the sun we approach the heavier metals, which are also compressed to enormous densities. The level between the heavy gases and light metals could constitute the perimeter of the eventual white dwarf stage of our sun.

White dwarf stars are very small, generally about the size of the earth. They constitute about 10 percent of the stars within close proximity of the sun. Their density is so enormous that one cubic centimeter will weigh about 900 kilograms, or one teaspoonful will weigh about five tons. Our sun will eventually become a white dwarf when it burns away the outer layer of gases. The level between the light metals and heavy metals could constitute the perimeter of the neutron star stage of our sun.

Neutron stars are much smaller than the white dwarfs. They are about twenty kilometers in diameter. That is about twelve and a half miles. Since they are so small and also radiate practically no light, they are virtually impossible to detect. Neutron stars are much denser than the white dwarfs. One cubic centimeter of this material could weigh about 200 billion kilograms, or one teaspoonful would weigh about a billion tons. Our sun will also eventually become a neutron star although it will be much smaller than the average. This will happen after the sun burns or erodes away through the white dwarf stage.

In the very beginning, during the formation of the sun, the central core was probably composed of iron and any other heavy metals in existence in the surrounding space at that time. This phase will be elaborated on in chapter 4, on the origin of the Milky Way and other galaxies.

The surface gas of the sun is at the extremely high temperature of about 6000 degrees centigrade. Now, to maintain a gas at such an enormously high temperature and at the same time keep it in a liquid state requires a tremendous amount of gravitational force.

Anyone familiar with the operation of an air compressor knows of the heat generated in the cylinders by the compression of the air. This heat has to be removed from around those cylinders by the movement of air by a fan or the pistons would expand and seize because of the heat. This is also true when any gases are liquified for industrial purposes such as oxygen, carbon dioxide, et cetera. These liquid gases are extremely cold because of the fact that so much heat was removed from them during the compression. They have to be contained in a tank at great pressure, otherwise they would rapidly evaporate. The gases on the sun, however, have no outer shell to contain them. Only gravitational force contains them.

During the original buildup and formation of the sun, the heat had nowhere to go because of the high vacuum surrounding it. Everyone is familiar with a vacuum bottle's ability to keep liquids hot or cold for hours. This is because the inner bottle is separated from the outer bottle by a vacuum. Subsequently, the heat during the sun's buildup rose until it reached the temperature at which it began to emit light. Since light readily goes through a vacuum, the sun then had an outlet for the intense heat generated.

The light leaving the sun seems to be in the form of particles known as *photons*. Do not confuse this with the *proton,* which is the nucleus of the hydrogen atom. The symbol identifying the photon is a short, wavy line (∿). This symbol signifies that they have the characteristics of waves and are considered to be packets of energy with no mass. I will explain this later on, in chapter 7. At the end of this chapter, I will also explain why the photon is really a particle rather than a wave.

These photonic particles leave the sun with such force that they drive away any gaseous substance near it. In that sense, they overcome the gravitational pull of the sun. An example of this pressure is the way the tails of comets always point away from the sun regardless of whether the comet is approaching or leaving the vicinity of the sun. The tails of the comets are probably composed of nothing but gases and very light particles.

Although science claims that photonic particles have no mass, I believe that they do have mass. But because of their extremely small size and enormous velocity, I'm inclined to think that that is why it is difficult to measure.

The velocity of the photons of light has been measured, however, and it is around 300,000 kilometers per second. That is the equivalent of 186,000 miles per second.

If light is a particle then it must have mass. The velocity of light slows down when it goes through a denser medium such as glass or water. This is a characteristic of particles. Waves, on the contrary, increase their velocity as they go through a denser medium. This is true of sound waves, and I'm sure it would be true of any other waves.

The photonic particles are probably formed from the collision of electrons. Therefore, they would have negative charges of electricity. A small percentage of protons within the central core of the sun would also collide to form some positively charged photonic particles. A more detailed description of atomic particles is given in chapter 7.

Because these particles are so small, the loss of mass by the sun is not large from year to year—4.26 billion kilograms of its matter is transformed into energy every second. That is the equivalent of 4.7 million tons. This may seem like a large amount, but when compared to the total mass of the sun, which is 2×10^{30} kilograms, it is very small indeed. Since it's difficult to express that sum in words, I've used the scientific way of expressing very large numbers. The exponent after the 10 means that 30 zeros follow the 2. That last figure is the equivalent of 2.2×10^{27} tons. That would be 22 followed by 26 zeros. In this instance, 27 figures follow the decimal point with the decimal point eliminated.

As the light from the sun is radiated into space, the gases will be used up first. Then the sun will be in the white dwarf stage. The radiation in this stage will be gamma rays, beta particles, alpha particles similar to the decay of radium, and a small quantity of ordinary light. These metals, however, as the result of their compression, are of much greater density than the radioactive metals on earth. So you can see why this material would be unstable—the surrounding gases are no longer there to keep them in a compressed and confined state. The gamma rays, beta particles, and alpha particles are associated with radioactive decay. The gamma rays are similar to X rays but have a shorter wavelength. To be specific, the shortest wavelength of the entire electromagnetic spectrum, of which visible light occupies a very small portion. Beta rays are ejected electrons. Alpha particles are helium nuclei.

This radiation will continue until the white dwarf stage of the sun erodes to the neutron star stage. During this stage the emissions will constitute neutrons, protons, and gamma radiation with secondary emissions of electrons and protons as the neutrons disintegrate. Outside of the nucleus, the neutron is an unstable particle with a lifespan of about fifteen minutes. With the continued emission of this neutron star stage, the sun will then completely disintegrate into nothing. Then the only remains will be the scattered particles of matter throughout space.

The reason I believe the neutron stars disintegrate is because of the free protons trapped among the neutrons as the shattered electrons are radiated from the center to the outer surface and into space during the compression of the core. With a small percentage of free protons in the core, the electrostatic repulsion between them would be great enough to force the disintegration, since the electrostatic force is billions upon billions of times greater than the gravitational force holding them together. Again, chapter 7 will have a more detailed explanation of these forces.

In the meantime, while the sun is disintegrating, Jupiter and Saturn will sweep the surrounding space of any free gases and acquire more mass. In the process, they will begin to radiate some light. However, they will never come close to acquiring the mass that the sun had. But they could acquire enough to radiate some red light. There was a recent report that Jupiter was already radiating a small quantity of red light.

If we were to hypothesize still further, say five or six billion years from today, we would see the sun continuously disintegrating in size. The planets would be acquiring more mass and their orbital radii around the sun would be increasing in size. Jupiter and Saturn would end up revolving around each other as miniature red dwarf stars. A body such as the planet earth could exist forever provided it did not collide with any other body, because it does not radiate any energy and can only acquire mass instead of losing it.

TYPES OF STARS

The stars radiate light according to the size of their mass. The smallest would radiate red light. If one of these smallest of stars was surrounded by a huge gaseous cloud it is unable to liquify because of its limited mass, it would be a giant red star. But it would still have a small mass because the central solid and partially liquified core would be comparable to a small star. Next in size would be the yellow, then the blue, and finally the giant blue-white stars. The giant blue and blue-white stars disintegrate at a much more rapid rate than the cool red stars. This rapid loss of mass is due to their enormous size and the resultant gravitational force that creates the extremely high temperatures these stars have. Even though they have much greater mass, they will have a shorter duration of existence than the cool red stars.

Once the stars reach the point of radiating light they will begin to lose mass, and if the solid central core is compressed sufficiently, it too will radiate after all the gaseous substance above it has been radiated into space. As mentioned previously, the stars will disintegrate completely. The particles that result from these disintegrations will form new atoms and molecules of gas to be reborn again into new stars and galaxies, as will be explained in chapter 4.

There are some who say that atomic energy is the only source of the sun's energy. But I'm inclined to think that the close crowding of the atoms in the core from the weight of the substances above causes the electrons to collide and shatter, with an occasional proton doing likewise. Because of the squeezing of the atoms, the electrons are pushed closer and closer to the protons, and in so doing, their velocities increase greatly. As their velocities increase, they lose some of their electric charge. This will be explained further in chapter 7. Once the particles collide, the cohesive bond holding them together is broken and the tremendous electrostatic force separates the particles like a violent explosion, so violent that the fragments are driven apart at the velocity of light. Of course, the cohesive bond holding the particles together under normal circumstances would be greater than the electrostatic force trying to separate them. It takes a collision to fragment the particle and break the cohesive force.

THE SUN'S MASS AND LIFE

If we were to take into consideration the total amount of the sun's mass destroyed in ten billion years, it would be 4.26×10^9 kg $\times$ 3.1536×10^{17} seconds. That last figure is the number of seconds in that portion of time. The product would be 1.3434×10^{27} kg. That's .00067 percent of the sun's total mass, which is less than one thousandth of one percent. If the present mass-loss continued at that rate until total disintegration, the sun would have a life span of 1.488×10^{13} years. That's ten thousand billion years. This is far in excess of what I believe modern science would have estimated the life of the sun to be. A more reasonable estimate would be about ten to forty billion years.

During those ten billion years, if we estimated that ten percent of the total mass destroyed was electron mass and the remainder proton mass, then 204 electrons would be destroyed for every proton. I would presume that would be about right because the electrons have a much higher velocity than the protons due to their mass differential, since the proton is 1,836 times heavier than the electron. Where the atoms would be squeezed closely together, the electrons would have velocities of more than one half the speed of light, while the protons would have much slower velocities. There would be a sufficient number of electrons left to form a considerable amount of neutrons and helium nuclei. And a number of free protons would be left over to effect the disintegration of the neutron star stage.

MASS OF PHOTON

It seems to me that the de Broglie formula cculd be used to calculate the mass of the photon. The formula states a relationship between the mass of a particle, its velocity and wavelength, as follows:

$$\lambda = \frac{h}{mv}$$

The wavelength is equal to the division of Planck's constant by the product of the mass times the velocity. The symbol for Planck's constant is h, while the wavelength symbol is λ. If we rearrange the formula to find the mass, we have:

$$m = \frac{h}{\lambda v}$$

To find the mass of a photon with a wavelength of $4{,}300 \times 10^{-10}$m in the blue light portion of the spectrum, we then substitute the following values: $h = 6.6262 \times 10^{-34}$ J/s, $\lambda = 4300 \times 10^{-10}$m, and $v = 2.998 \times 10^{8}$ m/s. The resultant mass then is 5.14×10^{-36} kgs. That means it would take about 177,000 photons of that wavelength to equal the mass of an electron. Is it any wonder, then, that these particles move at such a tremendous velocity?

MICHELSON AND MORLEY EXPERIMENT

What I consider further proof that substantiates the particle nature of the photon is the famous Michelson and Morley experiment. It was set up to determine the presence of an ether or substance that the waves of light would use as a transmission vehicle. If there was an ether, then this experiment would detect it. The basis of the experiment was to split a beam of light into two perpendicular beams by a partial reflecting central mirror. The beams were then reflected back toward the central mirror by two outer mirrors and toward an observer. I suggest the reader get a physics book for reference and a more thorough understanding of this experiment. By observing the interference of the waves in two perpendicular directions, one could then determine by also rotating the experiment 90° whether the velocity of light varied and subsequently whether an ether existed. The result was that there was no detectable difference. The experiment was repeated six months later with the earth moving in the opposite direction and there was still no detectable difference. This can be easily explained with the particle theory but cannot be explained with the wave theory.

The following is an example of what I mean. Let's imagine we have an experimental platform where we can pulse a light at two mirrors set at 90° angles to each other and 300,000 km away. I know that this is impossible, but for the sake of simplicity that size is ideal. We have an observer next to the light source who will time the pulses on their round trip to each mirror. In each instance, it takes two seconds, which of course is what we would expect since the velocity of light is 300,000 km per second. The scientific notation for the velocity of light is the letter c.

Now we set the platform in motion at 300 km per second, and represent that with the letter v. Now one of the light paths will be in line with the platform's motion and the other will be crosswise to it. We now send out a pulse of light in the direction of motion and discover that it takes two seconds for the round trip. We then send a pulse to the other mirror which is crosswise and discover that it also takes two seconds to complete the round trip. The motion of the platform has made absolutely no difference in the time it took to complete the round trip in both directions. With the particle theory this can be explained as follows: The motion of the light pulse in line with the motion of the platform is moving at 300,000 km per second, plus the added momentum of the platforms movement since the light pulse is a particle and therefore would have momentum. The total velocity of the first half of the light pulse is then 300,300 km per second, which is $c + v$. This velocity is relative to space. The velocity relative to the platform, however, is still 300,000 km per second. Before that pulse reaches the mirror, the platform has moved 300 km, so the total distance for the first half of the pulse to travel is 300,300 km, which is equal to the velocity and that means that it takes one second for it to reach the mirror. During the reflective process, as the light reverses its direction, the light loses velocity relative to space because it is now moving in the opposite direction from the motion of that mirror and that reduces its momentum accordingly. Its relative velocity to the mirror is still 300,000 km per second, but the mirror is moving away from the reflected pulse at 300 km per second. So the pulse's reflected velocity relative to space has now been reduced to 299,700 km per second, which is $c - v$. In the meantime, the platform has moved 300 km toward the pulse's return trip, so the actual distance for the pulse to return to the starting point has been reduced to 299,700 km, which means that the return trip of the pulse also takes one second. The total time then is two seconds for the round trip.

Now we send a pulse of light to the mirror moving crosswise to the direction of motion. The motion of the platform gives this pulse a small added momentum also. However, its added momentum will be much less than the initial pulse to the other mirror. We'll signify this velocity with the letters av, for angular velocity. This velocity will be equal to $c^2 + v^2 = av^2$, which is equal to 300,000.15 km per second. The return velocity will also be the same. Now this velocity will be equal to the actual distance that this pulse travels to the mirror. Therefore, this pulse would take one second to reach the mirror and one second to return for a total of two seconds. So in each direction it takes one second for the respective pulses to reach their mirrors and one second to return to their points of origin, this being true whether the platform is standing still or moving, or in which direction the pulse is traveling. Refer to figure 1.

Now with the wave theory of light we have a different situation. Since the light is supposed to be a wave and transmitted by an ether or whatever other vehicle of transmission there is, the velocity is fixed at 300,000 km per second and cannot exceed that speed because a wave does not gain any added momentum.

With a stationary platform, then, the light pulses would reach each mirror in one second and return in one second. This would be the same as the particle theory.

Now, with a moving platform, the light pulse in line with the movement would have to travel an additional 300 km. So with the additional distance the total elapsed time for the first half of the pulse would be 1.001 seconds. The time for the return half would be .999 seconds. The total time would be two seconds. Although the total time is two seconds, in this case each half has a different time for its completion.

The pulse for the crosswise direction would take 1.0000005 seconds to complete for each half for a total time of 2.000001 seconds. This, I believe, would have been detected during the Michelson and Morley experiment. The type of discrepancy would differ of course because of the slight difference between the experiment and my example. It is obvious that a photon of light must be a particle, since it conforms to the Michelson and Morley experiment where there would be no difference in the velocity of light regardless of which direction it was measured or in which direction the motion was.

The wave nature of the photon particle is given in chapter 7.

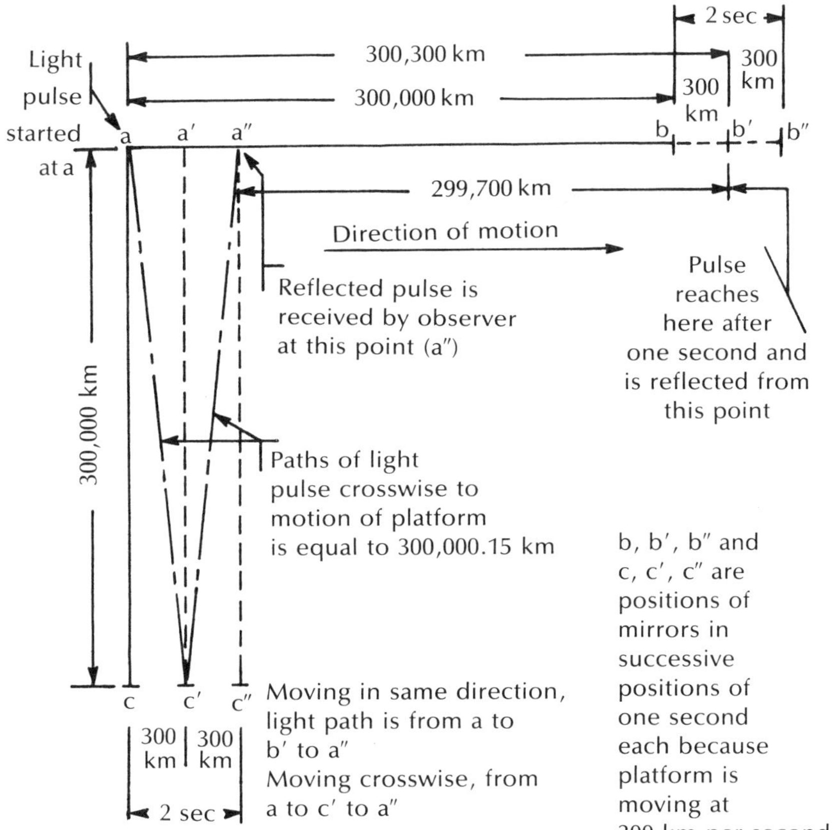

Paths of wave light pulses would be similar except the velocity would be limited to 300,000 km per second because the wave would have no added momentum.

Figure 1
Experiment with Light

Chapter 2

Sunspots and Related Objects

Sunspots are dark spots on the sun and are the remains of solar flares. According to present-day scientific opinion, sunspot activity is presumed to be caused by magnetic forces. However, there is no clear supposition in this respect, and there is no proof to substantiate this.

My belief is that sunspot activity results from the numerous meteoroids, asteroids, and comets that land on the sun, the solar flare resulting from the moment of impact of these bodies. These eruptions cause large quantities of liquid gas and liquified materials such as metals, silicon, and carbon to be thrown out into space. The remains of these bodies settle down to form the familiar spots.

The gravitational pull of Jupiter on these various bodies causes them to be directed into the sun. There are a large number of comets that approach the orbital region of Jupiter that are primarily affected by this.

When a meteoroid or asteroid lands on the sun, a single sunspot results. When a comet lands, a multiple sunspot results.

When these bodies enter the outer liquid shell of the sun, they rapidly melt and any gases within them are suddenly released. The oxygen component combines with the hydrogen to cause a violent explosion. These materials and gases are then blown out as far away as the orbits of Uranus and Neptune and possibly a distance beyond. However, the greatest amounts of these gases and solids are blown out to the orbits of Jupiter and Saturn. Because of this, the two planets have acquired the greatest amounts of these materials. That accounts for their massive sizes.

In the beginning, during the formation of our galaxy, the sun was probably a large blue star similar to the star Sirius in the constellation Canis Major. The star Sirius is actually a binary star composed of the main star, with its characteristic blue-white light, and a small companion, which is a white dwarf. However, in this instance, that is irrelevant since my reference is only to the main star.

During this time the planets were all hot molten cores of iron, silicon, and other solids and they were all about the same size. This includes the outer planets Jupiter, Saturn, Uranus, and Neptune. Of course, there were small variations in size such as that between the inner planets with the earth being two and a half times the diameter of

Mercury. However, there wasn't the enormous variations that exist between the inner and outer planets today.

Now, in the orbital region between Mars and Jupiter, there was in all probability two planets with one of them rotating in the opposite direction from the other. The inevitable result was that these two planets collided to form just about all the meteoroids, asteroids, and comets in our solar system. However, the collision didn't take place until all the planets cooled sufficiently to form solid bodies.

In all probability, the star Sirius has a planetary system also. This is probably true of the great majority of the stars in our galaxy and the universe. Although Sirius could have gained some mass from the companion star by accretion, I think its star system could be devoid of the small bodies that cause the star spot activity, and as a result it would have lost very little mass through that means. This would mean that its planetary system had no counter-rotating planets to collide with and produce these small bodies. On the other hand, our sun has appeared to have aged a lot more because of the great quantities of mass it lost as a result of the sunspot activity. But I'm sure Sirius is the same approximate age as our sun.

The solid materials blown out in the solar flares form spherical particles close to a millimeter in size as the result of the intense heat. Proof of the existence of these particles was found in the soil samples taken from the moon. I'm sure these particles came from the solar flare activity of the sun.

Although most of these particles are composed of silicon, there are some composed of other elements as well.

The remaining dense matter from the flare activity remains in the sun as a spot or spots until the heavier material slowly settles to the bottom of the liquid gaseous layer.

The particles that reach the earth cause the disturbances that interfere with shortwave radio communications as they enter our atmosphere. They also cause the auroras at the north and south poles. Further elaboration on the auroras will be given in chapter 3. They are also responsible for the superheated Van Allen radiation belts which encircle our earth. These will also be elaborated on in chapter 3.

The occurrence of spots on the sun varies in frequency from a minimum to a maximum in eleven years. That is, it averages out to a period of eleven years because the cycles vary slightly in duration. This is known as the solar activity cycle or the sunspot cycle. Refer to figure 2.

The cause of the variation is a heavier concentration of the elliptical orbits of the comets and other bodies in one region of the solar system. Where the concentration is the greatest, that would correspond to the

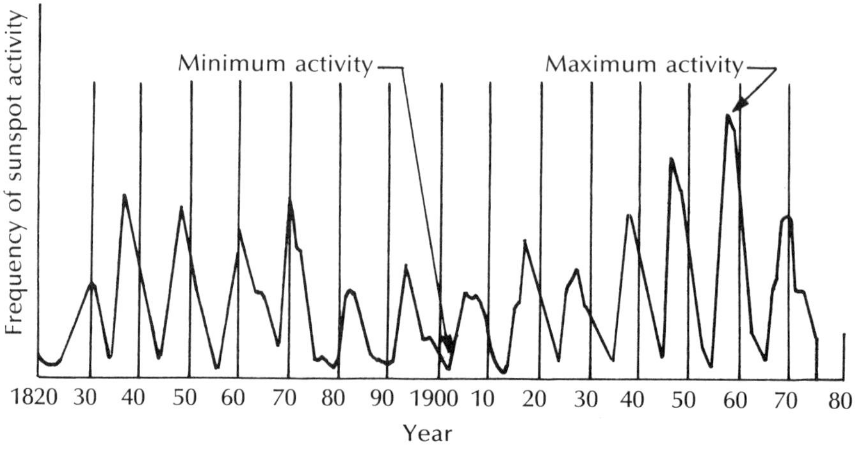

Variation of sunspot activity averages out to eleven-year period

Figure 2
Sunspot Activity Cycle

peak activity of the sunspot cycle while the region where the spots are sparse would correspond to the least activity, as illustrated in figure 3.

The orbit of the planet Jupiter is 11.86 years in duration, while the sunspot cycle is approximately 11 years. This discrepancy will no doubt be used by proponents of other theories to discredit my explanation of sunspots. However, I will ask, what happens when a meteoroid, asteroid, or comet does land on the sun? The obvious answer would be a solar flare and a resulting sunspot later. We know that these bodies do fall into the sun because the probability that they do is so great. All the planets and satellites in our solar system are cratered by these bodies and the sun would be no exception. Because of the heat of the sun, the reaction is different.

The vast majority of the comets and other bodies have an elliptical revolutionary period of six to seven years. The aphelions of these orbits reach out to Jupiter's orbit. The aphelion is the point of the orbit that is the farthest away from the sun. The gravitational pull of the planet causes a disruption in their movement and it results in the flight of some into the path of the sun.

Practically all of these bodies have elongated elliptical orbits with the major axis slowly revolving in the opposite direction to that of the planets. The major axis is the line drawn through the center of the perihelion and the aphelion and runs lengthwise through the elliptical orbit. The perihelion is the point of the orbit closest to the sun. The

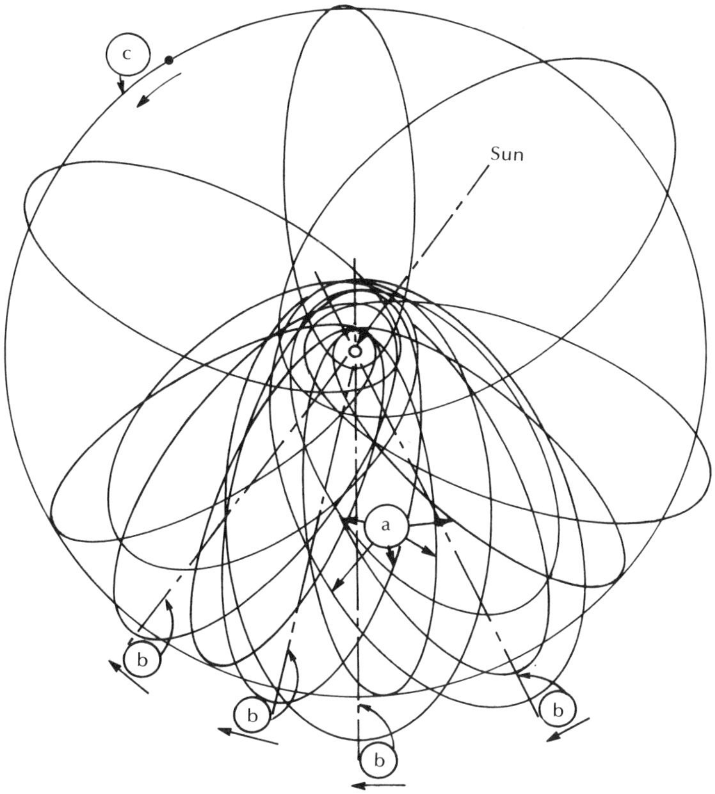

Figure 3
Cause of Sunspot Cycle

Text explains significance of letters

combined movement of all these axes of all these bodies counter to Jupiter's rotation shortens the sunspot cycle to eleven years as illustrated in figure 3.

The explanation of the significance of the letters in figure 3 is as follows:

(**a**) You'll notice that the arrows point to the orbits of the various comets in that region. They are also concentrated in that one region of the solar system to account for the variation in sunspot activity. This concentration corresponds to the maximum activity of the sunspot cycle.

(**b**) Arrows point to the major axis of the comet orbits, while the bottom arrows point to the direction of movement of the axis which is in a clockwise direction. This movement is approximately 26° per year, which shortens the sunspot cycle to eleven years relative to Jupiter's period of 11.86 years. Jupiter's disturbance of the orbits of these various bodies at its own orbit directs a number of them into the sun.

(**c**) Jupiter in its orbit with arrow indicating its movement in a counterclockwise direction, contrary to the movement of the major axis of the comets.

NOVAS AND SUPERNOVAS

A nova appears to be an exploding star that rapidly increases in magnitude. It may increase on the average by as much as ten magnitudes. That is an increase by as much as 10,000 times in brightness.

However, a nova is nothing but a gigantic star flare that probably was caused by the impact of a small planetary body the size of Pluto. With such an impact, the star could lose a sizeable portion of its mass, depending on the type of material of which the planet is made.

Supernovas also appear to be exploding stars and their luminosity may increase by more than twenty magnitudes. That's an increase in luminosity of close to a billion times. Contrary to present-day astronomical thinking, a supernova is not a collapsing star and is not an atomic explosion that is supposed to be the result of that collapse. Supernovas are caused by the same thing a nova is caused by except that the impacting planetary body may be much larger and of lesser density. A body of lesser density is composed of gaseous substances primarily and so the force of the explosion would be much greater than a dense planetary body would cause.

The position and direction that a planetary body lands on a star would determine whether the resulting nebula caused from the star flare would be uniform or irregular. Where the impacting body lands in the center and at a perpendicular angle, the nebula would be generally uniform and symmetrical. If the impacting body lands at an angle, the resulting nebula could be irregular. The degree of irregularity would depend on the acuteness of the angle at which the impacting body lands.

The depth of penetration by the impacting body as the result of its density, size, and composition would also have a bearing on the size of the explosion and the amount of stellar mass blown away. It would also have an effect on the type of nebula that would result.

Three factors, then, are involved in the size and configuration of the novas and supernovas. They are: size of impacting body, angle of entry, and the density or composition of the impacting body.

The Ring Nebula M57 in the constellation Lyra is an example of a central and perpendicular impact by a sizeable planetary body of lighter

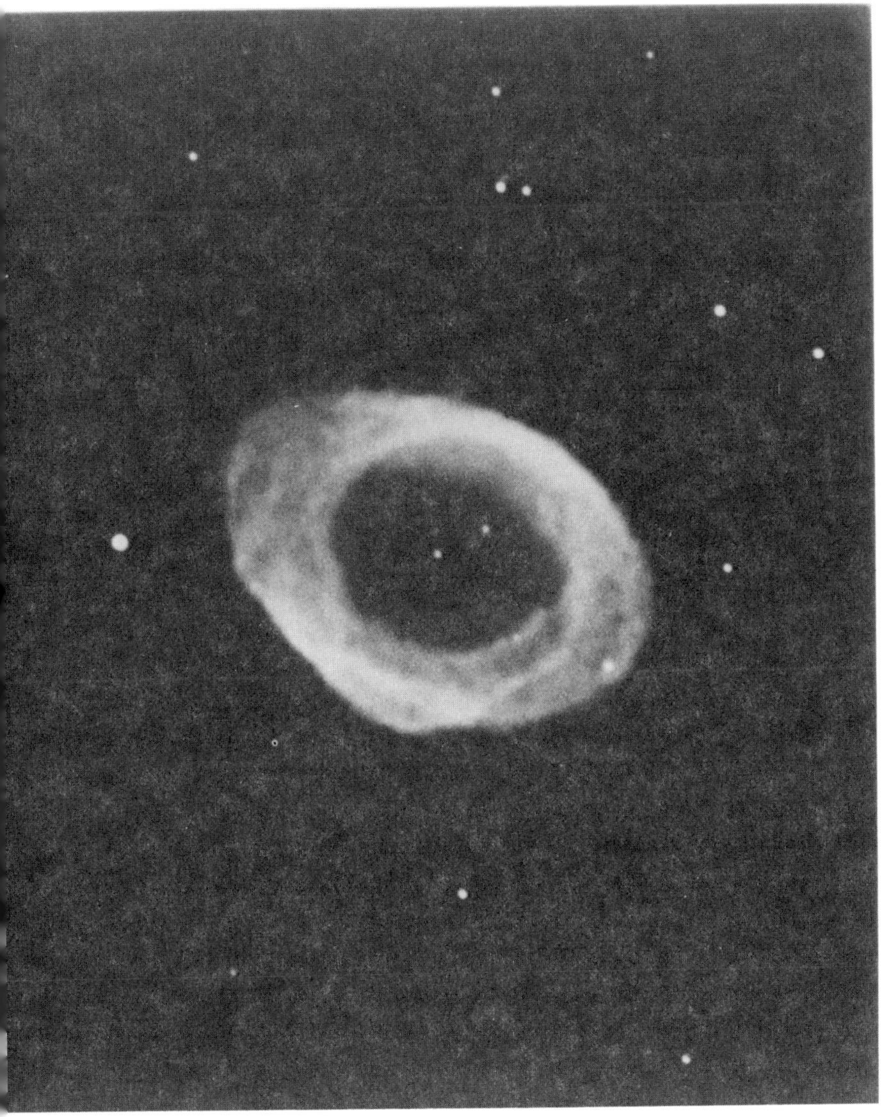

Figure 4
Ring Nebula in the Constellation Lyra (NGC 6720, M57)
(© Palomar Observatory)

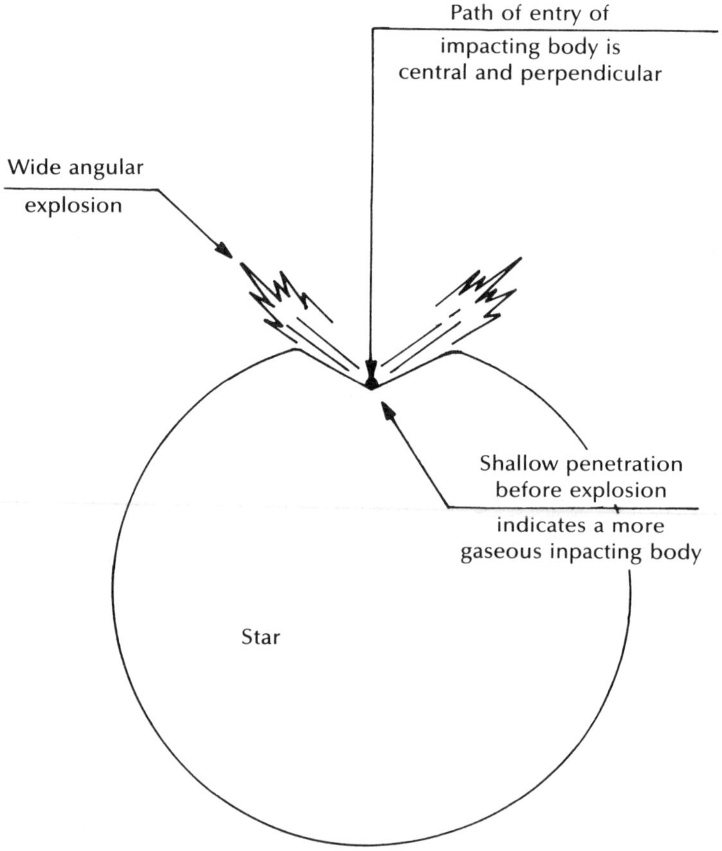

Figure 4a
Possible Type of Explosion for Ring Nebula and NGC 7293

density. I believe the explosion took place close to the surface of the star, which resulted in a more widespread explosion rather than one directed back toward the direction of the impacting body. Refer to figures 4 and 4a. Also the outer diameter of the ring relative to the diameter of the star would indicate that the nova is older rather than more recent. Some of these planetary novas can be up to 50,000 years old.

Another example of this type of planetary nebula is NGC 7293 in the constellation Aquarius. However, this nebula was probably caused by a more massive impacting body because of the larger quantity of

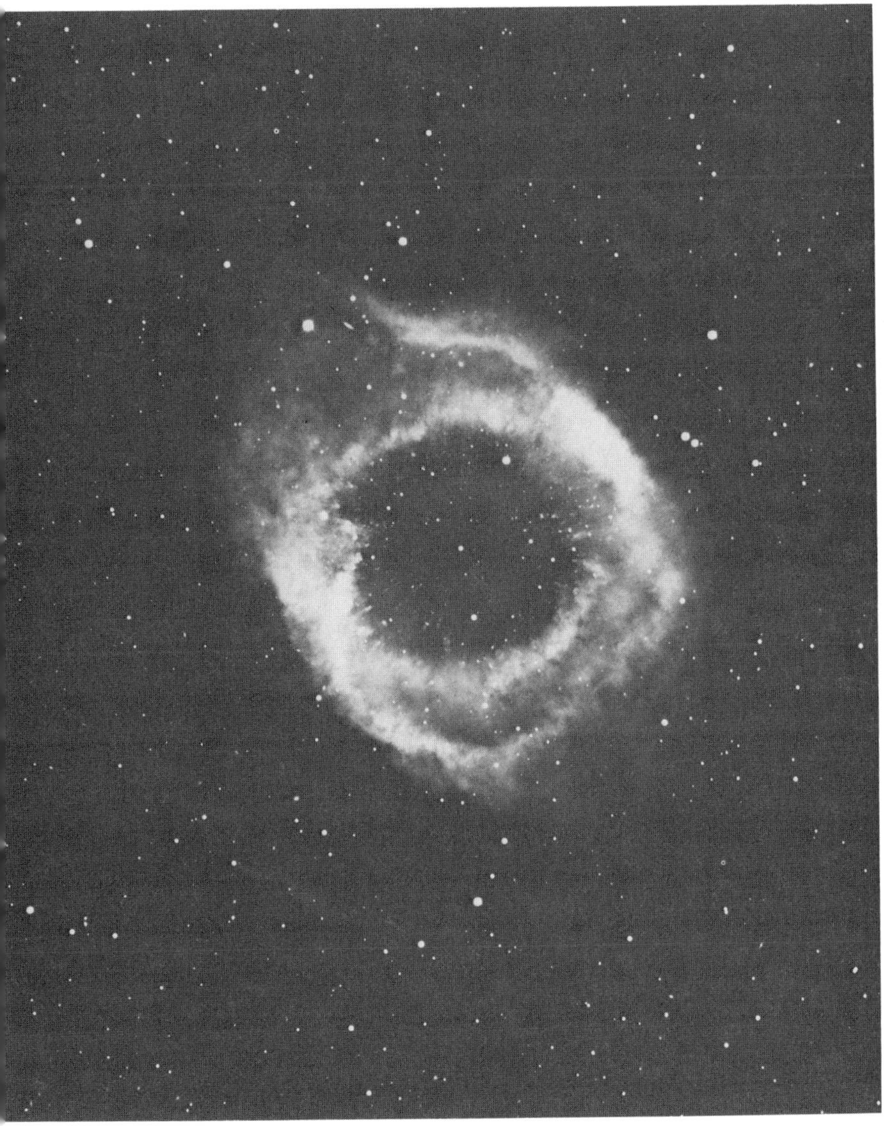

Figure 5
The Helix Nebula in Aquarius (NGC 7293)
(© 1973 AURA, Inc. Kitt Peak National Observatory)

gas blown out. The star appears to be much smaller relative to th nebula and could have lost a considerable quantity of gas. Refer t figure 5.

The Owl Nebula M97 in the constellation Ursa Major is also a nov resulting from a central and perpendicular impact by a planetary body However, the penetration here could have been much deeper befor the explosion happened and the result is that the gases appear to b blown out more acutely backward. Refer to figures 6 and 6a. Thi explosion is also not as old as the Ring Nebula because the size of th outer perimeter of the nebula relative to the size of the star's diamete in the center is not as great as the Ring Nebula's.

Another nebula similar to the Owl Nebula is NGC 2392, which i sometimes referred to as the Eskimo Nebula. This nebula is in th constellation Gemini. It appears to be much younger than the Ow Nebula because the relative difference between the diameter of th star and the outer perimeter of the nebula is much less. Also the oute perimeter of the central portion hasn't had as much time to disperse Refer to figure 7.

The Dumbbell Nebula M27 in the constellation Vulpecula is an example of an irregular. This nebula resulted no doubt from a planetary body passing through the edge of the star at an acute angle and also exiting at the back edge, creating a somewhat symmetrical pattern Refer to figures 8 and 8a.

Most nebulas are irregular and are nothing but concentrations o gas in the spiral arms of the Milky Way. They are not the result o explosions but are the remains of gases that were not condensed during the formation of the galaxy and, to a lesser extent, random gases blown out during star spot activity. These gases left over from the original formation just haven't been condensed because they were out of reach of the stars during their formation. These gases are present in other spiral galaxies as well. There will be more explained about this in chapter 4.

THE CRAB NEBULA

The Crab Nebula in the constellation Taurus is a supernova remnant from an explosion that was observed by the Chinese in A.D. 1054. It is an irregular-shaped mass of gas blown out from an angular penetration by a very massive planetary body, probably from two to five times the mass of the earth. The star in this case could have lost from 25 to 50 percent of its gaseous mass but gained a sizeable quantity of denser solid mass in return.

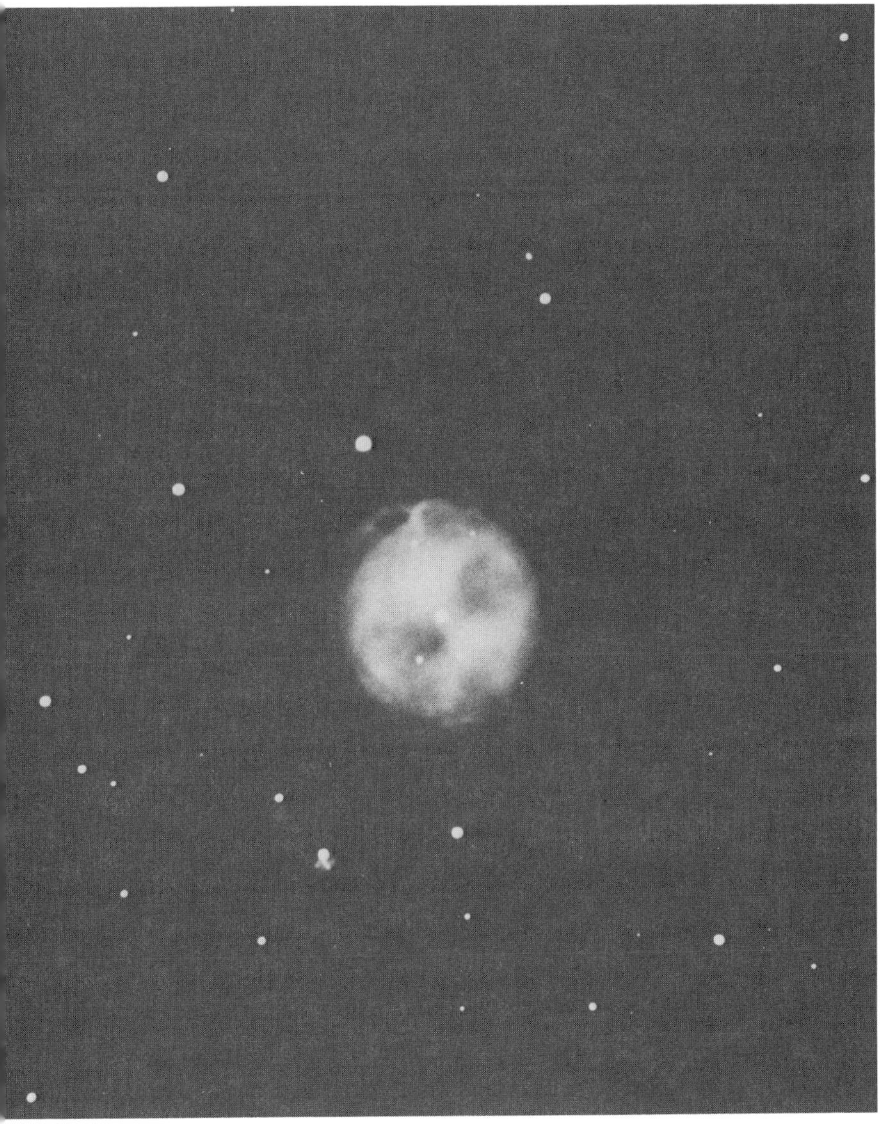

Figure 6
Owl Nebula in Ursa Major (NGC 3587, M97)
(© Mount Wilson and Las Campanas Observatories,
Carnegie Institute of Washington)

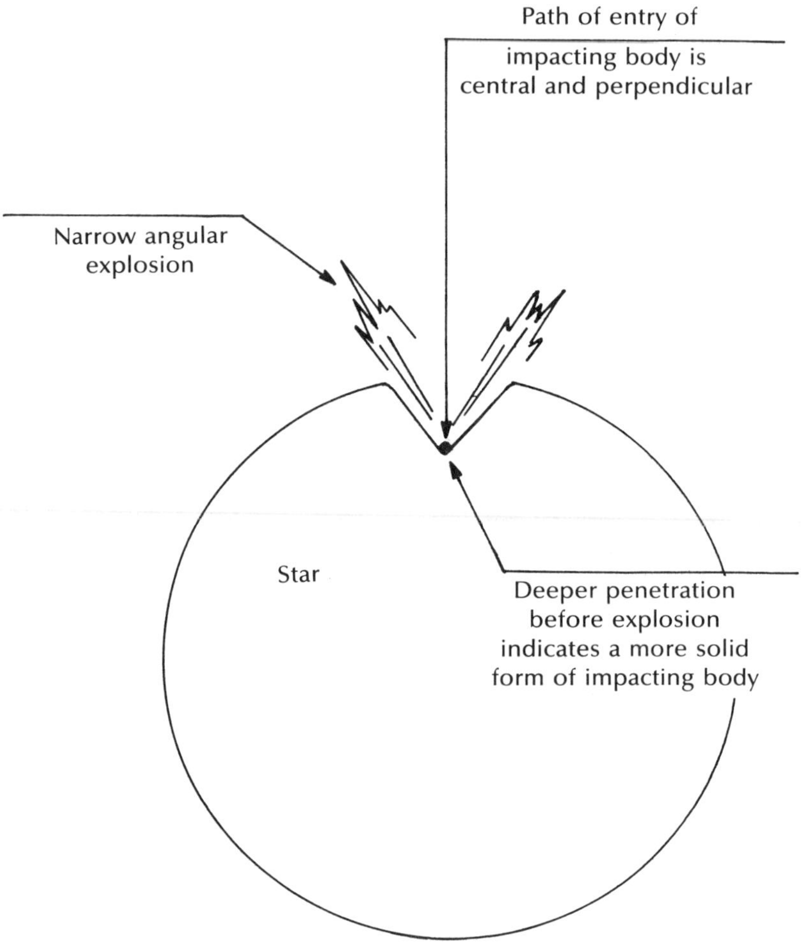

Figure 6a
Possible Type of Explosion for Owl Nebula and NGC 2392

The gas seems to have been blown out in our direction and a star in the center has been suspected as the source of this supernova explosion. The star has a continuous spectrum that lacks absorption lines which indicates that it is spinning at a high rate of speed.

By means of radio astronomy, which uses radio telescopes instead of the conventional visible-light telescopes, a pulsar was located close to the central region of the nebula. A pulsar is an object in the sky that emits steady periodic bursts of radio waves. This particular pulsar has a frequency of thirty times a second which would indicate that a

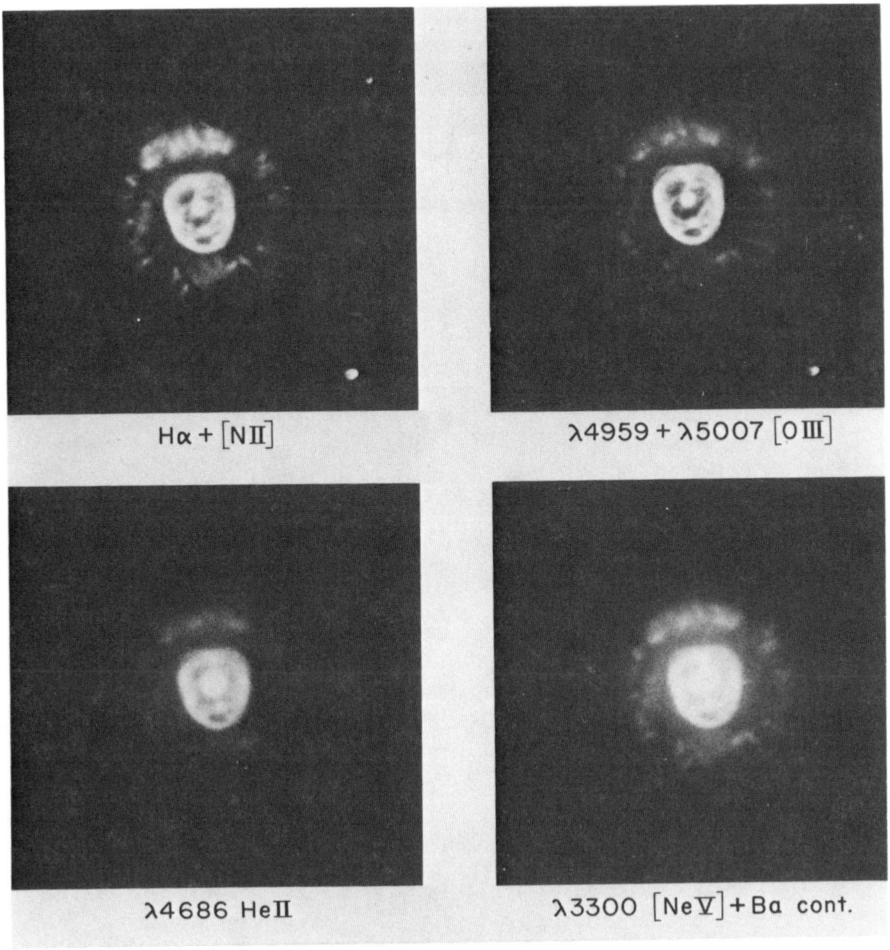

Figure 7
NGC 2392 Planetary Nebula in Gemini
(© Palomar Observatory)

star is spinning at that high rate. It could also mean that the mass of the impacting body separated into two halves on opposite sides for balance. If that were true, then the spin rate would be reduced to fifteen times a second as will be explained later. This pulsar is known as Taurus A.

The suspected star was then observed through a stroboscopic device that permitted observation through a slit in a disc rotating at thirty revolutions per second. It was then established that this star did blink

Figure 8
The Dumbell Nebula in Vulpecula (NGC 6853 and M27)
(© Kitt Peak National Observatory, Cerro Tololo Inter-American Observatory)

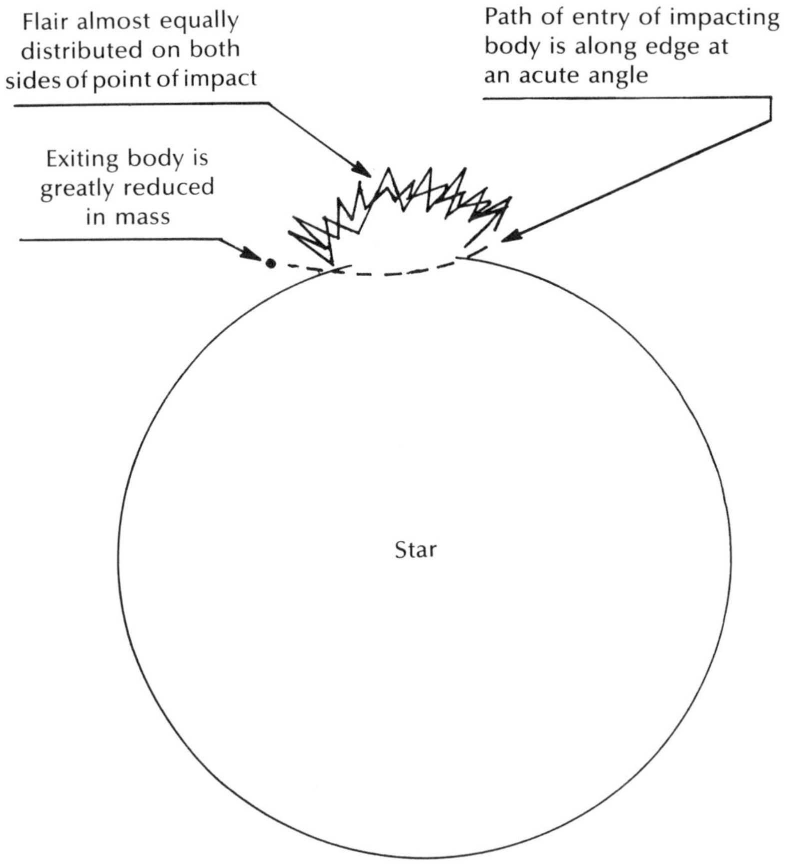

Figure 8a
Possible Type of Explosion for Dumbbell Nebula

at the same frequency of thirty times per second as the pulsar's frequency. It was then determined to be the source of the supernova.

Because this star is presumed to be spinning at the high rate of thirty times a second, the assumption is that this is a neutron star. Since supernovas are considered to be nuclear explosions from within the star, then all the gaseous substance is blown away and the only thing remaining in its place is a neutron star. Since a neutron star is only 20 km in diameter, it's pretty obvious that no neutron stars have been observed as yet. Also, the high spinning velocity can only be withstood by a neutron star, because of its high gravitational force and small diameter. Other stars would tear apart because of their large diameters and the large centrifugal forces generated at those high spinning velocities.

I contend that it is an ordinary star, as I've explained previously, because it is plainly visible. Neutron stars are not. The high spinning velocity was caused by the explosion of the planetary body at great depth; also because it entered at an angle, and that is the most important reason why the star is spinning at such a high velocity. I also believe the spin rate is fifteen rather then thirty times a second. According to an observation by the Einstein X-ray satellite (HEAO 2), the X-ray pulses from the Crab Pulsar alternate in intensity, which would indicate that the impacting body split in two, each half taking positions on opposite sides of the star. This would then reduce the spin rate to one-half, or fifteen times a second, the force of the explosion from the impacting body being directed to the side and tangent wise to the center of the star. Refer to figures 9 and 9a.

What holds the star together? The strong gravitational force of the central region and a strong magnetic force field by the residue of the impacting body could hold the star together. The free protons in the central core and the free electrons in the residue could also contribute in holding the star together.

The irregularity in the star's nebula is also an indication that it was not a nuclear explosion. That type of explosion would have caused a uniform-spherical gas nebula since the gases would have been directed evenly in all directions.

The pulses, of course, are the result of two gigantic star spots on opposite sides of the star from the split of the planetary body. Since the sunspots on our sun cause great quantities of radio emissions and magnetic storm disturbances, plus the darkening of the sun in those regions, it seems reasonable to conclude that the same situation exists on the star of the Crab Nebula. One great difference, however, is the size of the impacting body and the resultant spots. In this situation we have two spots that cover about a quarter of the star's surface on opposite sides, and also the fact that they have remained that large for so long a time. The persistence of the spots over their long life span of 925 years is due to their gigantic size and the high rotational velocity of the star. The spots remain surfaced all this time because of their greater density and resulting greater centrifugal force relative to the liquid gaseous substance around them. These spots will continue to remain surfaced as long as the star continues to spin at its high rate of speed.

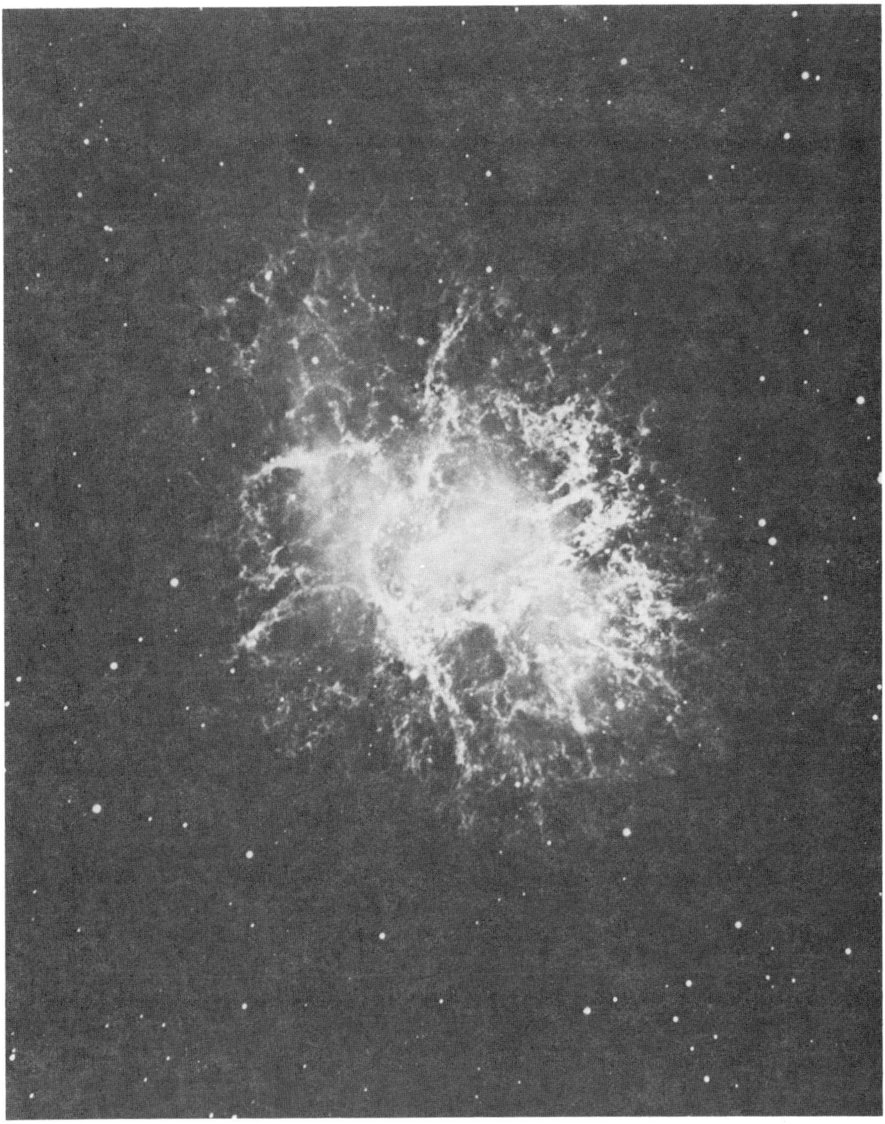

Figure 9
The Crab Nebula in Taurus (NGC 1952 and M1)
(© California Institute of Technology and Carnegie Institution)

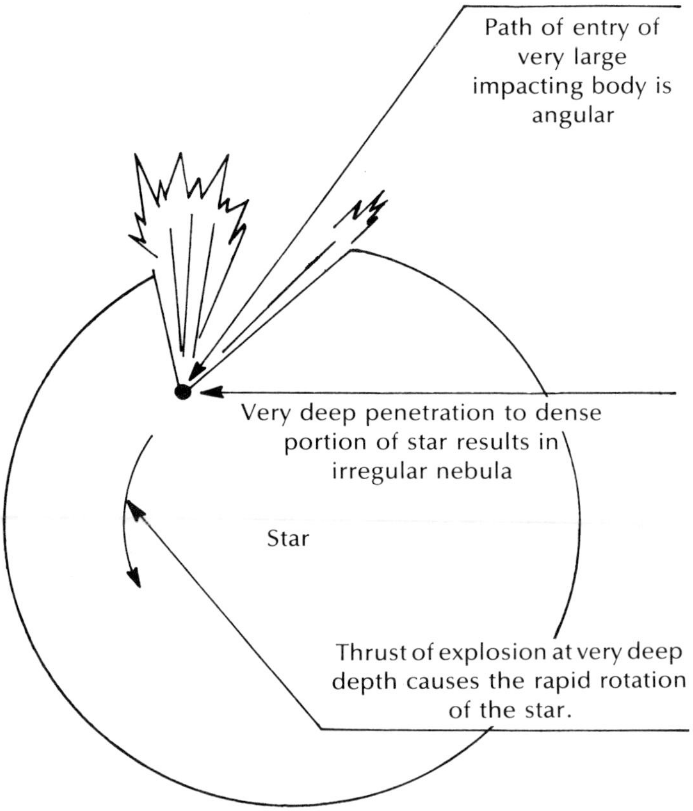

Figure 9a
Possible Type of Explosion for the Crab Nebula

CEPHEID VARIABLES

The Cepheid variable stars are divided into two classifications known as Type I and Type II Cepheids. The Type I Cepheid is also known as the Classical Cepheid, and this is the one I'll be concerned with here, although I still would like to explain the difference between the two classes.

The Type I Cepheid is simply the brighter of the two classes by about 1.5 magnitudes. However, the Type I Cepheids are located in different galaxies, while the Type II Cepheids are located primarily in the globular clusters which surround our Milky Way galaxy. The variable periods of both classes vary from about one to a hundred days and are very regular.

The Type I or Classical Cepheid is supposed to be a supergiant pulsating star where the physical size is supposedly expanding and contracting with the luminosity varying accordingly. These stars are thousands of times brighter than our sun. An example is a star in our own galaxy known as δ Cephei (delta of Cepheus) near the constellation Cepheus. Its magnitude varies in 5.4 days from 3.6 to 4.3. It is also the first star of this type discovered.

These stars are primarily used to measure the distance of galaxies because of the close relationship between their luminosity and period. The brighter stars have the longer periods. Since their magnitude can be determined to some degree, though not with great accuracy, by the length of their periods, they can be used to measure distance. By assigning them an absolute magnitude taken from a table that is associated with their period and by observing their apparent magnitude (visual), one can determine how far away they are.

The absolute magnitude of a star is the brightness it would have if placed at a fixed distance of ten parsecs, the parsec being a unit of measurement in astronomy that is equal to 3.086×10^{15} km or 3.26 light years. The reader should also be familiar with the fact that the magnitude scale is reversed. The higher the magnitude number, the lower the luminosity of the star. The brightest stars in the heavens have negative numbers.

However, my concern here is not their use in measuring distance but the cause of their pulsations. The pulsations are not caused by the expanding or contracting of the star. The Cepheid Variable is simply a binary star but of a slightly different nature. The ordinary binary has two stars with one generally darker than the other. When the darker one obscures the brighter one either partly or wholly while they orbit around each other, you then have a typical binary of variable luminosity. Of course, the plane of their orbit would have to be in line with our line of vision. With the plane of the orbit perpendicular, there would be no variation.

In the Cepheid Variable, the smaller star is orbiting the larger star in an orbit that is elliptical and off-center. The smaller orbiting body could be a white dwarf, a neutron star, or possibly a planetary body, the orbit being sufficiently off-center to cause contact between the two when the smaller body skips along the outer perimeter of the larger star during its perihelion passage. This causes a star flare of a various size depending on how deeply the smaller body touches the surface. The frequent and repetitious flares form a gigantic gaseous shell around both bodies and that gives the appearance of a supergiant star. As the flares erupt, they light up the whole outer shell. The brightness of the shell to our line of vision would depend on which side the flares would be erupting. (Refer to figure 10.)

Since the flare would be driven out rapidly and fall back slowly, the wave pattern of the light variation would correspond to the characteristic light patterns of the Cepheid Variables. (See figure 10a.)

These pulsations or flares would continue for thousands or millions of years, depending on the erosion of the smaller body. A planetary body may last for thousands of years but a white dwarf or a neutron star could cause the flares to last for millions of years, because there would be very little erosion from either, because of their extremely dense material. Eventually the smaller body would lose velocity to a point where it would fall into the larger star and never emerge again. At this time it could cause one last gigantic flare to form a nova. This could cause the giant star to lose a sizeable percentage of its mass or very little depending on the gas-content release of the falling body. The planetary body could cause a large explosion, the white dwarf a lesser one, and the neutron star the least of all. The increased mass from the latter two could cause the regular star to brighten due to the increased mass.

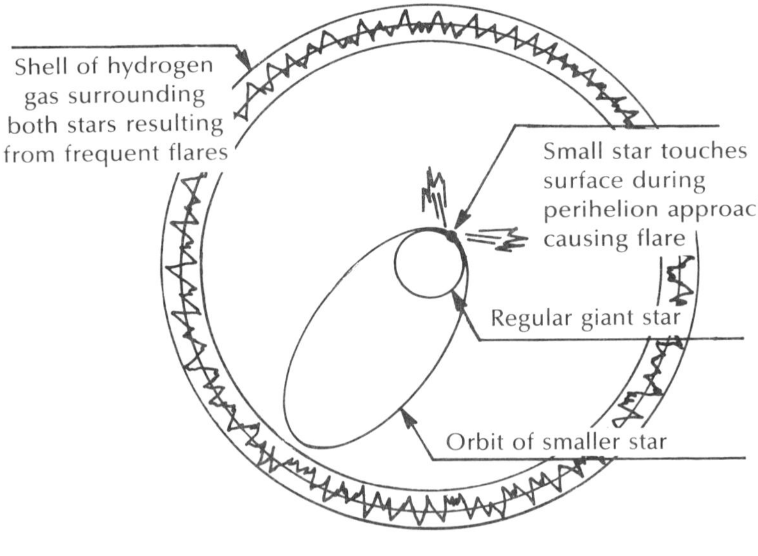

Figure 10
Cepheid Variable

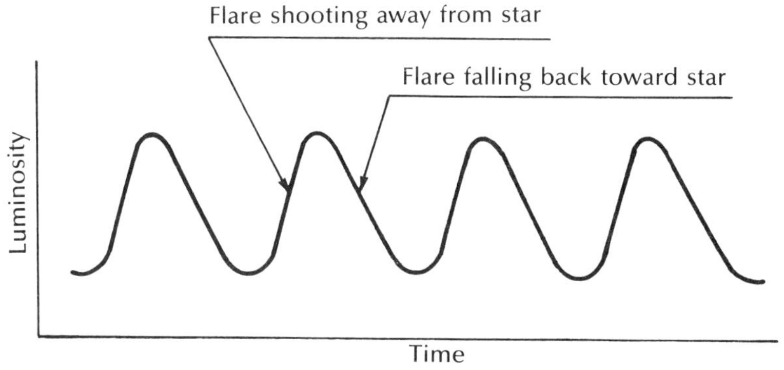

Figure 10a
Wave pattern of Cepheid Variable

CYGNUS A

Cygnus A is a source of powerful radio waves. The source was just recently pinpointed by the most advanced of radio telescopes. When the giant two-hundred-inch telescope at Mount Palomar photographed that region, it was discovered that a peculiar galaxy was located at that point. It looks as though two elliptical galaxies have collided, the distorted shapes of each resulting from that collision.

The radio source comes from two lobes of radiation on opposite sides of the optical source. (Refer to figures 11 and 11a.) The radio lobes are obviously the emitting regions of a gigantic ring nebula that surrounds the two galaxies. Since the plane of the ring is in line with our vision, it appears as a lobe on each side. The radio telescope is receiving radiation from the longest linear densities of the ring. (Refer to figure 11b.)

This is another example of a collision between heavenly bodies. However, in this instance, it is not just two bodies colliding such as in a nova or supernova, but a great number of stars in each galaxy colliding to form a gigantic gaseous ring nebula. The galaxies are not at present colliding even though they appear to be doing so. The collision has already occurred during their first passage. They cannot be in collision at the present time because no gaseous streamers are visible, and because the ring is already at such a great distance from the galaxies. They are locked in an elliptical orbit and are in the present positions of their orbits that places them in our line of sight, creating the illusion that they are together. No doubt, though, they are separated by a sufficient distance to avoid any further collisions.

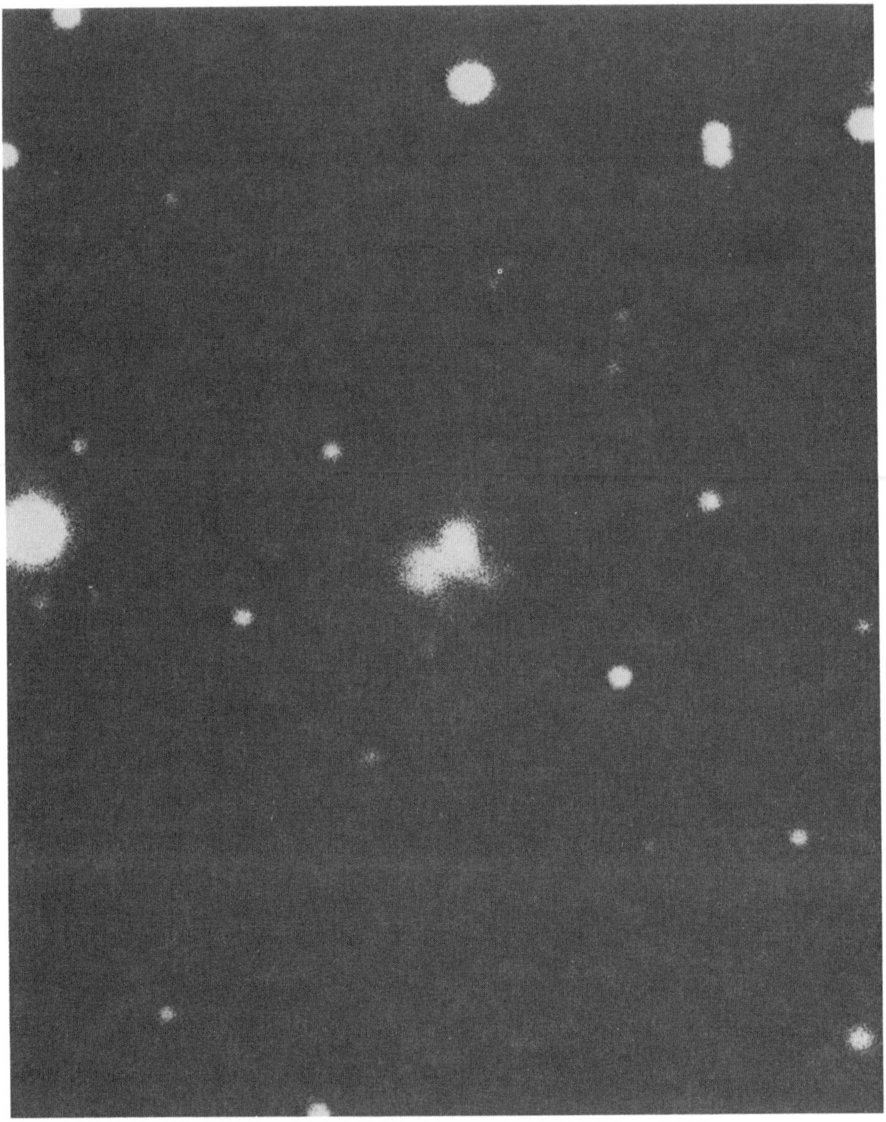

Figure 11
Cygnus A Radio Source
(© Palomar Observatory)

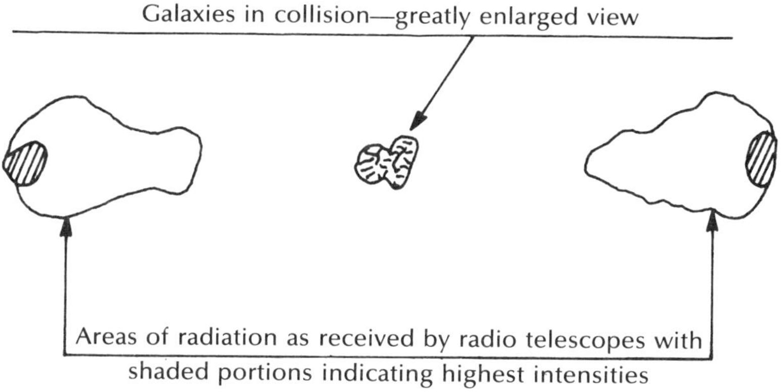

Figure 11a
Cygnus A Radio Lobes

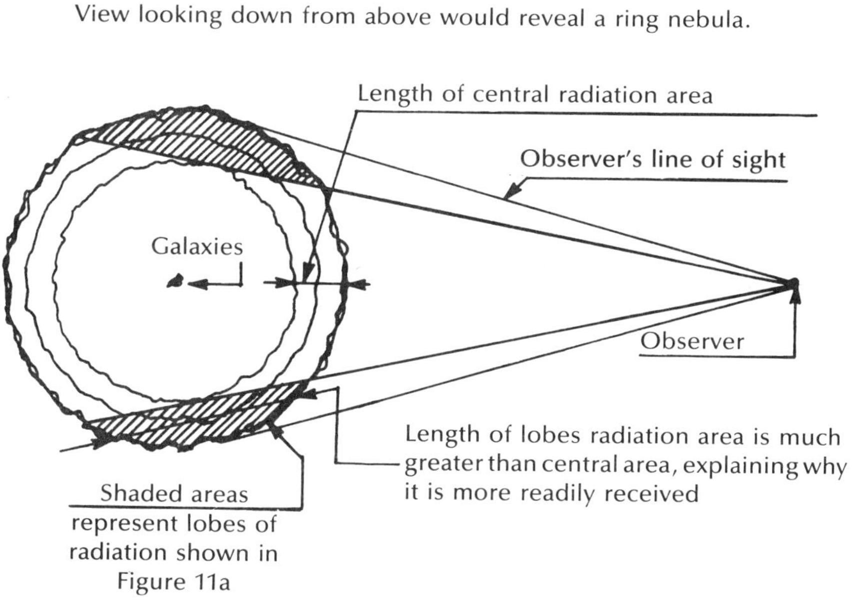

Figure 11b
Cygnus A as Ring Nebula

The flat side of the right elliptical galaxy was the side involved in the initial encounter, and since the galaxies are locked in orbit and possibly rotating, then that would explain why the flat side is on the opposite side of the present apparent collision. The other galaxy in all probability also has a flat side, but it too has rotated. The flat side could be on the front or back side at present, which would mean that it rotated 90° after the initial impact with the rotation vertical and perpendicular to our line of vision. These two galaxies could have completed one or two orbits since the initial encounter.

All radio galaxies having two lobes on each side of the visible source are gigantic ring nebulas. The causes, of course, are similar to Cygnus A. However, this does not mean that two galaxies are always involved although the exceptions would constitute a small percentage of the total. Stars can collide within a galaxy and a sufficient number of these collisions could add up to cause these radio nebulas to form, such as in Centaurus A. (Figure 20 in chapter 4 is an optical image of Centaurus A. Radio lobes are not shown.)

Other radio sources are caused by streamers of gases directed in one direction only from the galaxy. An example would be the elliptical galaxy M87 in Virgo. (Refer to figure 12.) This also is caused by a star collision. Probably a white dwarf with a regular star, in the central core of the galaxy, where no doubt collisions occur frequently because of the high concentration of stars in the nucleus. The white dwarf's deep penetration and lack of gaseous molecules would probably cause an explosion directly back toward the line of entry, resulting in a concentrated streamer towards the outside of the galaxy.

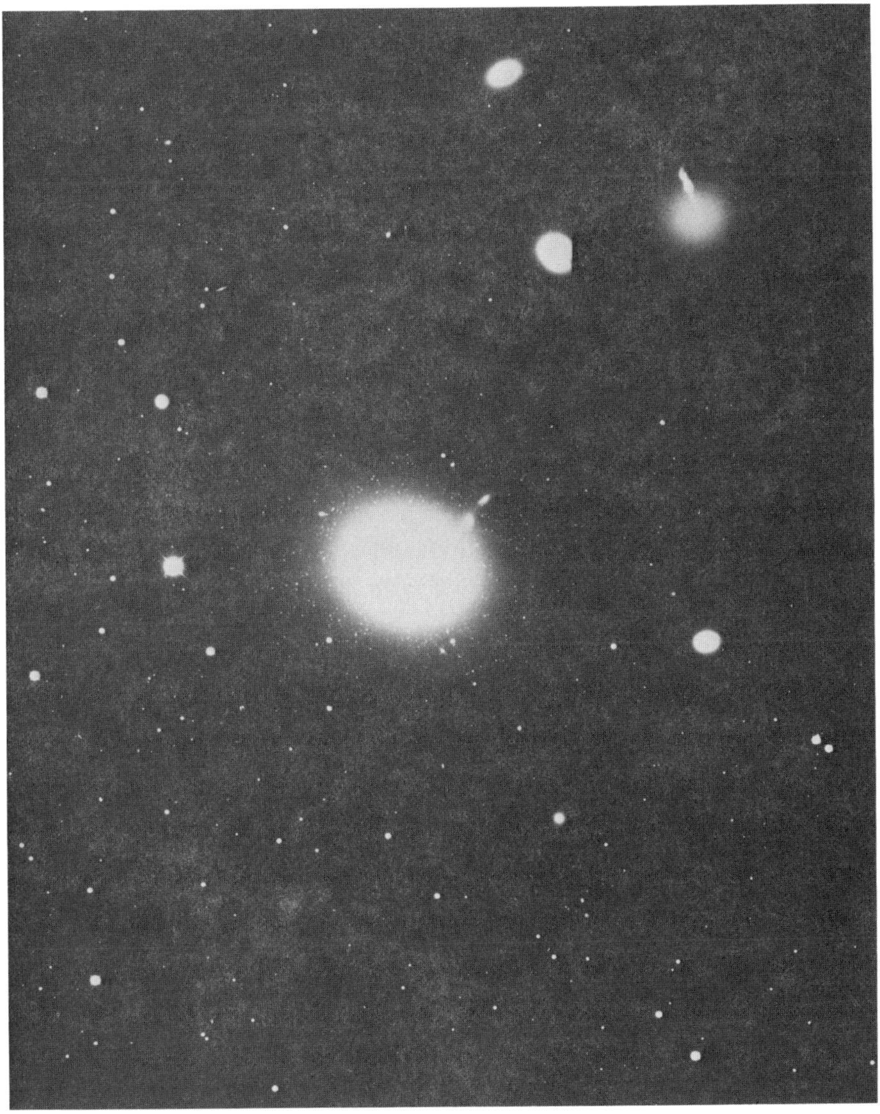

Figure 12
Galaxy M87 (NGC 4486) in Virgo, Virgo A Radio Source.
(© Kitt Peak National Observatory, Cerro Tololo Inter-American Observatory)

Chapter 3

The Aurora Borealis and Van Allen Radiation Belts

The aurora borealis, commonly known as the northern lights, are composed of a variety of different forms and colors near the North Pole. Similar lights are also seen in the southern hemisphere and are known as the aurora australis. Since both auroras are the result of the same causes, this discussion will be devoted to the aurora borealis.

The aurora borealis is partly caused by the particles from the solar flare eruptions on the sun. However, there is another and greater contributing factor to the auroras. This is the solar flare particles that are combined with the meteorite particles that are left from the disintegration of the meteors during their entry into the upper limits of our lower atmosphere. These particles settle in the (e) and (f) layers of the atmosphere surrounding our earth.

Our atmosphere seems to be composed predominantly of four layers of gases. The outermost hydrogen layer has an altitude of about 20,000 km to 27,000 km above the earth at the equator. The second layer is helium at an altitude of about 2,000 km to 5,000 km, also at the equator. These two gases constitute the upper portion of our atmosphere and do not mix with the lower portion to any great extent. Their relationship to the Van Allen radiation belts will be discussed in the latter part of this chapter.

The third and fourth layers are the two primary gases of the lower part of our atmosphere. They are the nitrogen and oxygen components, and I believe these two gases form the (e) and (f) layers. The (e) layer is the upper level of the oxygen component and is approximately 100 km above the earth, while the (f) layer is the upper level of the nitrogen component and is about 160 km high.

The (e) and (f) layers have been designated as such because of their ability to reflect the short radio wavelengths that are used in radio communication for amateur and commercial purposes.

The oxygen and nitrogen layers settle at their respective heights because of their different densities just as two liquids of different densities would settle in a bottle (this happening only when there is no disturbance). These smooth, settled surfaces plus the disintegrated meteorite and solar flare particles I've mentioned previously combine to act as reflectors.

However, during solar flare activity, the flare particles from the sun penetrate through the hydrogen and helium layers to these two surfaces and cause them to literally boil because of the heat generated by friction. The result is that the smooth surfaces are disturbed and the resultant turbulence can no longer reflect the radio waves. This then results in a temporary halt of radio communications. The two disturbed layers can then combine to form a single thick layer as the result of the turbulence.

The auroras result from the movement of the atmosphere containing the (e) and (f) layers to the polar regions, the movement depending on the weather conditions at the surface of the earth. These levels move in toward their respective polar regions as a single or a double layer of particles depending on whether they were in a tranquil or a turbulent state.

This upper air movement toward the poles is the result of the cold air at ground level moving away from the poles. This cold ground air replaces any air that has risen at the lower latitudes and tropical regions as a result of the sun's heat. If you have watched the weather forecasting programs, you will notice that every now and then when the right pressure patterns exist, cold air moves down from Canada that originally came from the polar regions and moves over the United States. This air moves down all the way to Florida and the rest of the southern states. A similar situation also occurs in Europe, Russia, and Asia.

Some of the explorers to the North Pole have reported the heavily dust-laden air in that region. There were also reports that open water existed at the North Pole. This would be highly unusual because of the extreme cold that exists there. Only the warm air coming down from above can account for this.

At the southern hemisphere the polar region is completely covered with ice because of the lack of any large land areas in close proximity. Consequently, the volume of air descending there is not as great as at the North Pole.

The northern lights occur as single and dual bands, streamers, arcs, and as a faint glow at the uppermost levels.

As the dust-laden air of the (e) and (f) layers moves down vertically from the upper level to the lower level and the tranquility of the layers has not been disturbed, then they obviously would appear as two bands of light. This light being generated by electricity between the particles coming from the flares on the sun and the particles of the (e) and (f) layers at the poles, the particles being electrically charged from the friction as they entered the atmosphere. Some of this light could result from reflection and refraction of the sun's light also. This reflected or refracted light could be the result of the transparent spheroidal silicon particles. The carbon or iron particles could also reflect some light even though they would be completely black. This would be because their surfaces would be smooth and shiny. The electricity would account for most of the light, though. Nitrogen and oxygen ions account for most of the color although refracted light from the silicon particles could account for some. (See figure 13.)

Of course when the (e) and (f) layers are intermixed as a result of former turbulence en route to the North Pole, then a single wavy band will result. This situation exists more often then the dual bands. The single bands are long streaks that sometimes fold back in various patterns. (See figure 14.)

The streamers seen in the northern lights are the undisturbed particles' streaks from the numerous meteorites that are seen as meteors disintegrating in the region of the (e) and (f) layers. Whether they are at the same level as the layers or below, they have remained intact from turbulence to the time they enter the polar region. At times these streamers seem to be converging to one point as they move in from all directions. These streamers may have resulted from a meteor shower that the earth had encountered previously. (See figures 15 and 15a.)

The arcs are a single layer of dust entering the polar region while the dust is still approaching in its horizontal position before it begins its descent to the surface of the pole. (See figure 16.)

The faint glow at the highest level of the northern lights is probably helium gas which is at a distance above the (e) and (f) layers. This glow resulting from the solar flare particles entering and generating a considerable amount of electricity.

THE VAN ALLEN RADIATION BELTS

The Van Allen radiation belts are superheated regions in space high above the earth and are mostly concentrated in the equatorial region. The reason for the existence of the belts has not yet been clearly evaluated by present-day science. However, their existence can be explained as follows:

The belts are nothing but superheated hydrogen and helium gases. These layers of gases surrounding the earth at two different heights, as I've mentioned before, the lower belt being the helium and the upper being the hydrogen belt. These gases are very rare so close to a perfect vacuum, but they do exist there. Since these gases are very light, the

Figure 13
Aurora Borealis Dual Bands

Figure 14
Aurora Borealis Single Band

Figure 15
Aurora Borealis Coronal Rayed Band

Figure 15a
Aurora Borealis Coronal Rayed Band

Figure 16
Aurora Borealis Arc

gravitational pull of the earth is not sufficiently strong to bring them any closer. But the earth's gravity does manage to contain them at those very high altitudes. They are unevenly distributed around the earth because of the variation of their temperatures and resultant density. This creates oval-shaped layers instead of spherical ones. The rare density also prevents the heat from dispersing evenly throughout these gases. The heat, of course, is generated by the solar flare particles and meteorites that penetrate these gases, as I've mentioned before.

The heat is concentrated in the equatorial regions because the earth's gravity draws the particles toward the central or equatorial zone of the atmosphere. This is especially true of the lower belt.

The inner helium zone is a small oval around the central portion of the earth while the upper hydrogen layer has a crescent-shaped radiation zone. The space temperature in between and around the radiation zones falls off rapidly to very low levels as the distance increases away from the zones. Figure 17 is a cross-sectional illustration of the zones.

As the particles from the sun and outer space penetrate the gases, their velocity slows down and so their trajectories are curved toward

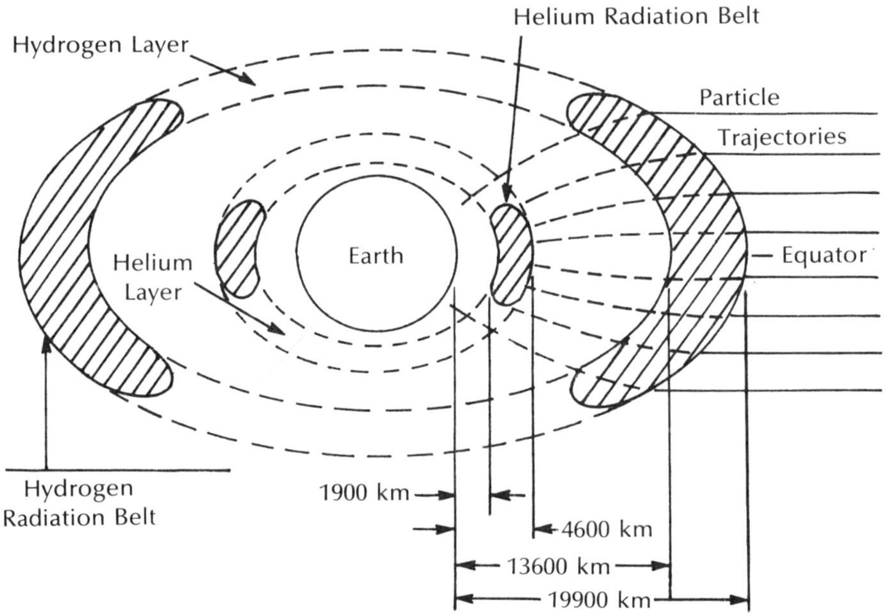

Figure 17
Van Allen Radiation Belts

the center of the earth. Since the equatorial regions are at a much higher temperature than the polar regions, the layers are thicker there due to the expansion of the gases. Also, they are at a higher altitude due to the resultant lighter densities. That, then, is what causes the oval-shaped zones. The combined oval shape of the layers and the trajectory of the particles is what creates the shape of the radiation zones.

The very lightest of particles may disintegrate in the hydrogen layer. Then, as the particles vary in size and density, their penetration will vary accordingly. The heaviest of particles penetrating to a depth below the (e) and (f) layers of the earth's lower atmosphere. Of course, the paths of the trajectories will also vary in relation to the density of the particles.

Chapter 4

The Perpetual Universe

How do you define the word *universe*? My definition would be: All matter and space within the confines of a fixed region in total space including the light itself. Excluded from the definition would be the space beyond our material universe because space is infinite in its dimensions. All space outside our universe would be totally devoid of matter and light. The light leaving our universe would not continue out to infinity because the gravitational pull of the combined mass of the universe would cause the light to eventually stop and fall back. Gravity has an effect on light because, as I've mentioned in chapter 1, light is a photonic particle of matter. It has already been proven that the gravitational pull of stars does bend light.

THE NATURE OF THE UNIVERSE

How did our universe originate? According to present-day astronomical science, the universe originated with a huge explosion commonly known as "the big bang." I'm inclined to think, however, that a great number of our astronomers do not accept the big bang account. I also cannot accept it because it leaves too many important questions unanswered.

What caused the big bang? Matter does not explode spontaneously. How did the matter get congregated in the center of the universe in the first place? Matter in space never seems to congregate in the center due to the motion of the bodies. Their momentum gives them a tendency to continue in a straight line. I'm sure everyone is familiar with this from riding in a car rounding a curve. That is known as centrifugal force. In that respect those bodies resist gravity or act as a negative gravitational force. Gravity is the force that imparts motion but the bodies seem to end up revolving around each other.

What happened before the big bang? Where did the matter come from? These and other questions pertaining to the big bang theory will never be answered because the theory itself does not sound realistic.

On the other hand, the steady-state theory does sound realistic because the galaxies and stars, like life itself, constantly renew themselves. There is no beginning or end for the universe. With this concept, those questions pertaining to the big bang are either eliminated or resolved.

THE ORIGIN OF GALAXIES AND STARS

As I've mentioned in the first chapter, stars disintegrate into tiny particles of matter until nothing is left, or more accurately, into tiny particles of matter with electrostatic or electromagnetic properties. This disintegration results in the radiation of light, radio waves, X rays, and gamma rays by the photons, electrons, alpha particles, protons, and neutrons that are being ejected. The stars, then, are not converted into pure energy. These tiny particles go on to form new atoms and molecules in faraway intergalactic space. An example would be a free proton that attracts a free electron to form a hydrogen atom or a sufficient number of photons to build up its mass to that of a neutron. In this way photons would be regenerated back to electrons because the neutron is an unstable particle outside the nucleus of an atom.

The formation of the new gases from the ejected particles then could be held together by their weak gravitational force to form gigantic clouds in between the regions of the galaxies. Since the gaseous atoms and molecules are considered to be electrically neutral, their weak gravitational force would be the only thing to form the clouds and hold them together. Because of this, the densities of these clouds would be very rare and, as a result, would be enormous in size. After a considerable period of time, however, they would capture and accumulate a considerable amount of dust and particles of matter to increase their densities. With this increased density, they could then accumulate larger bodies even up to planetary size. This increased density would reduce their overall size to some extent. Since these gases and bodies would all have momentum, this would keep the cloud from condensing into a compact cloud or mass. The gravitational force would then hold the clouds together while the momentum of the matter would prevent the cloud from condensing. If the cloud remained for a great period of time, it would evolve into a spherical cloud. A large number of the younger clouds would exist as elongated streaks of matter as a galaxy ejects its matter on its journey through space.

But how did the Milky Way galaxy, the sun, and the planets originate in the first place? To begin with, the entire solar system was formed at the same time approximately as the Milky Way galaxy.

The old galaxies create new galaxies. As a giant galaxy moves through space ejecting particles in its path, it will eventually enter the neutron star stage. By this, I mean it will have a majority of neutron stars. It will be in this stage that the galaxy will have the power to create a new galaxy. With a majority of neutron stars it could be referred to as an old galaxy. With all these neutron stars ejecting tremendous amounts of electrons, protons and nuetrons, there would also be produced a tremendous amount of X rays and gamma rays. The motion of these particles through any cloud they enter would produce a lot of secondary radiation in the form of radio microwaves and subsequent heat from these microwaves. It is all this radiation and heat that has a lot to do with new galaxy and star formation. An old galaxy in this stage would in all probability be invisible because there would be very few stars radiating visible light.

Now with a galaxy of this type approaching an intergalactic cloud of matter composed of various compounds of gas, dust and larger bodies, the cloud would be superheated by all this radiation because it is very penetrating. It is because of this penetrating effect that a new galaxy can be formed. Everyone is familiar with the penetrating effect of X rays and radio microwaves while the gamma rays have the greatest penetrating power of all.

X rays are constantly used to photograph parts of the human body. They are also used to check for flaws in steel parts and other substances. Radio microwaves are used for fast cooking because they penetrate completely and heat the food throughout, all at the same time.

On the other hand, ordinary light does not have any penetrating power because it is easily reflected, and although it does go through a transparent or translucent substance, it does not generate much heat. Because of this, the newer galaxies which radiate primarily nothing but ordinary light cannot induce any heat in any intergalactic cloud that they pass. As a result, they cannot create any new galaxies.

The resultant heat induced by the old galaxy then causes all the dust and other solid compounds within the cloud to melt and expel all the gases contained within them. This is important because it is the first step in star formation. The heat also drives all the free electrons out of the solids.

This principle is similar to the electronic vacuum tube with the heated cathodes. When the cathode is heated, it emits electrons. Of course, the vacuum tubes are practically obsolete nowadays since the semiconductors have replaced them. Vacuum tubes are still used, however, for television picture tubes and video monitors for computers.

The expulsion of the electrons from the particles of matter causes them to attach themselves to the gas atoms and molecules. This produces negative gas ions. This emission could be a contributing factor to reinforcing the gravitational force all bodies possess. It is also a very important factor in the star buildup. This will be explained later.

The movement of the galaxy as it passes the cloud also imparts motion to all the matter within the cloud by its gravitational pull. The cloud would then flatten out and rotate in the direction that the old galaxy is moving. This would apply only in the formation of a new spiral galaxy when the old galaxy is at a distance.

The creation and formation of the various types of galaxies will be explained later.

The process of the formation of a new galaxy could take up to a billion years before the old galaxy was out of range to induce any more effective heat from its radiation on the assumption that the old galaxy is moving at approximately 800 km (500 miles) per second. The newly created galaxy's movement through space would be somewhat slower and would depend on its total mass and on how close it was to the passing galaxy.

THE SEPARATION AND BUILDUP

How do the bodies get to be so massive to form stars and planets?

The separation of the gases and solids by the intense heat generated by the radiation of the passing galaxy is the first important step in the formation of the stars.

We start out with a gigantic intergalactic space cloud of gas and tiny bits of dust and some larger bodies of matter scattered throughout. The movement imparted to the solid particles by the gravitational force of the passing galaxy causes the red-hot liquid solids to separate from the gases surrounding them, the solids having greater mass and momentum than the gases. The gases are moved but not as much the solids.

In each of the vast regions of space separated by, say, four to eight light years distant, the most massive body within that region will act as the central nucleus and all other smaller bodies will begin to revolve around it. However, within the totality of that particular region will be billions upon billions of tiny particles revolving around each other and other larger bodies nearby. In the meantime, the gases lacking momentum and mass begin to slowly drift in toward the central nucleus because of its larger mass and resultant slower velocity relative to that region in space. Each of the extremely large regions within the gigantic cloud acts like a giant centrifuge with the lightest substances converging toward the centers of their respective nucleus.

I'm sure everyone is familiar with the principle of separating substances of varying densities by the use of centrifugal force.

There will be a considerable number of particles and bodies that start their rotation in the wrong direction. By this I mean contrary to the rotation of the majority of the others. These, however, will have their movement retarded as they move against the motion of the gravitational pull of the passing galaxy and the gravitational pull of the other particles and bodies nearby. As a result, they will eventually lose their momentum and be drawn into their respective central region by merging with the central nucleus or with the largest body in their own planetary space. However, some star systems will have one or two planetary bodies still moving against the majority after the condensing and merging process has been completed, our own solar system being one as I've mentioned in chapter 2.

While the gases are converging toward the central nucleus, the tiny heated particles will be the first to lose their momentum and will merge with the larger bodies nearby while the gases are still being scattered throughout the surrounding regions as they are moving toward the central nucleus. The heavier bodies keep their momentum as they move through these gases. In the meantime they acquire more velocity as they acquire more mass by merging with other bodies moving in the same direction nearby, because the passing galaxy is still exerting its gravitational pull. As these superheated spherical bodies continue to merge and merge and grow and grow, they eventually reach planetary size. They will then end up as planetary bodies around the nucleus within their particular star region. There will be millions, and in the case of a giant galaxy, billions, of star regions.

In the meantime, the gases converging toward the molten central nuclei of their respective star regions and composed mostly of hydrogen atoms and molecules charged as negative ions are being drawn and condensed around the nuclei because of their gravitational pull and electron deficiencies. As I've mentioned before, the heated and molten bodies are all charged positive because the heat has driven off all the free electrons and some of the outer orbital electrons as well from their various atom components. This combined gravitational and electrical force continues to condense the gases more and more. The weight of the gases upon the central nucleus causes the temperature within the nucleus to rise still further. This in turn causes more electrons to move to the outer perimeter of the gaseous envelope. This causes a greater attraction between the electron-deficient nucleus and the outer electron-rich gaseous envelope. Further condensation then results as the gases continue to converge, the intense heat continually causing the outer electrons of the atoms to skip to the outer perimeter of the newly forming star. The weight of the buildup by the gases and their electrical attraction causes the growth of the star to continue until the central nucleus is crushed to enormous density, the density of a neutron star. Succeeding lesser densities of all the various elements within the new star continue all the way to the outer surface.

This process continues until all the gases are liquified and no further condensation can take place. With the completion of the condensation, the central core ceases to generate any more heat. Then it slowly begins to cool until it reaches a temperature of about one or two thousand degrees above that of the outer surface of the star. This central heat is dissipated by conduction within the metallic portion of the nucleus and by convection within the liquified gaseous portion of the star. The temperature is then finally stabilized at the surface and will be determined by the size of the star, the size of the star depending on the amount of gaseous matter within that particular star region of the new galaxy. This process is going on in each separate star region within the new galaxy. The planets take about the same amount of time to merge and grow as the star, and cease to grow any further in size when they reach a point where they are too far apart to merge any more. Each star system may not take exactly the same amount of time to develop but the time variation would not amount to very much in astronomical terms.

The final condensation of the stars and the merging of the molten particles and bodies to planetary size, as I've mentioned, may take up to a billion years to complete, the planetary bodies all being in a molten state at this point. Thus a new galaxy is complete and in existence.

After the completion, the planets will begin to cool—slowly, of course, with the lightest elements being at the surface of the newly formed planets. The cooling of the planets then permits their gravitational fields to attract some of the residual gases toward themselves to form atmospheres, and no doubt life, as they continue to cool further.

The stars, of course, continue to stay extremely hot and shine because of their gravitational force resulting from their huge size. However, the heat now is generated mainly from this gravitational force and a sizeable portion from the nuclear energy contributed from within the core.

The portion of the original cloud with the largest amount of total mass in it would become the center of the new galaxy with the greatest amount of star systems.

The portion of the newly formed galaxy nearest the old passing galaxy would probably form slightly ahead of the other side and have a slightly larger concentration of stars. The differential in time, however, would not be too great, possibly from twenty-five million to one hundred million years.

These newly formed galaxies would contain a large number of blue-white stars. These stars would remain blue-white for the longest periods of time if they have no counter-rotating planets, because, as I've mentioned earlier, counter-rotating planets cause collisions which produce asteroids, meteoroids, and comets. These in turn produce star flares similar to the sun's solar flares as they land on the stars and cause large quantities of gas to be blown out into the surrounding space. The resultant losses of gas reduces their size and eventually changes their color also.

As this new galaxy grows older, the stars would go through the various stages of aging. Eventually, when it reaches the old age of a galaxy rich in neutron stars, it too will be able to produce a new galaxy from some giant intergalactic cloud in space.

The oldest galaxies in the universe would be moving at the highest velocities because they are composed of extremely dense neutron and white dwarf star systems and because of the lack of gaseous substance between those star systems. This creates very little resistance to the scattered matter in space. The resistance of this matter would be very small. However, I think its effect on the velocity would have some significance.

CLASSIFICATION OF GALAXIES

Before I explain how the different types of galaxies are formed, I think the reader should be familiar with the different types and how they are classified.

Galaxies are mainly classified into four different groups. They are the ellipticals, spirals, barred spirals, and the irregulars. (Refer to figure 18.) There are other classifications pertaining to a small percentage of galaxies but we will not be concerned with them here.

The first group classification we need be concerned with is the ellipticals. These galaxies are spherical to long ellipticals in shape and are designated by the capital letter E. A numeral follows the letter to denote the elongation of the ellipse. Zero is for a galaxy that is spherical,

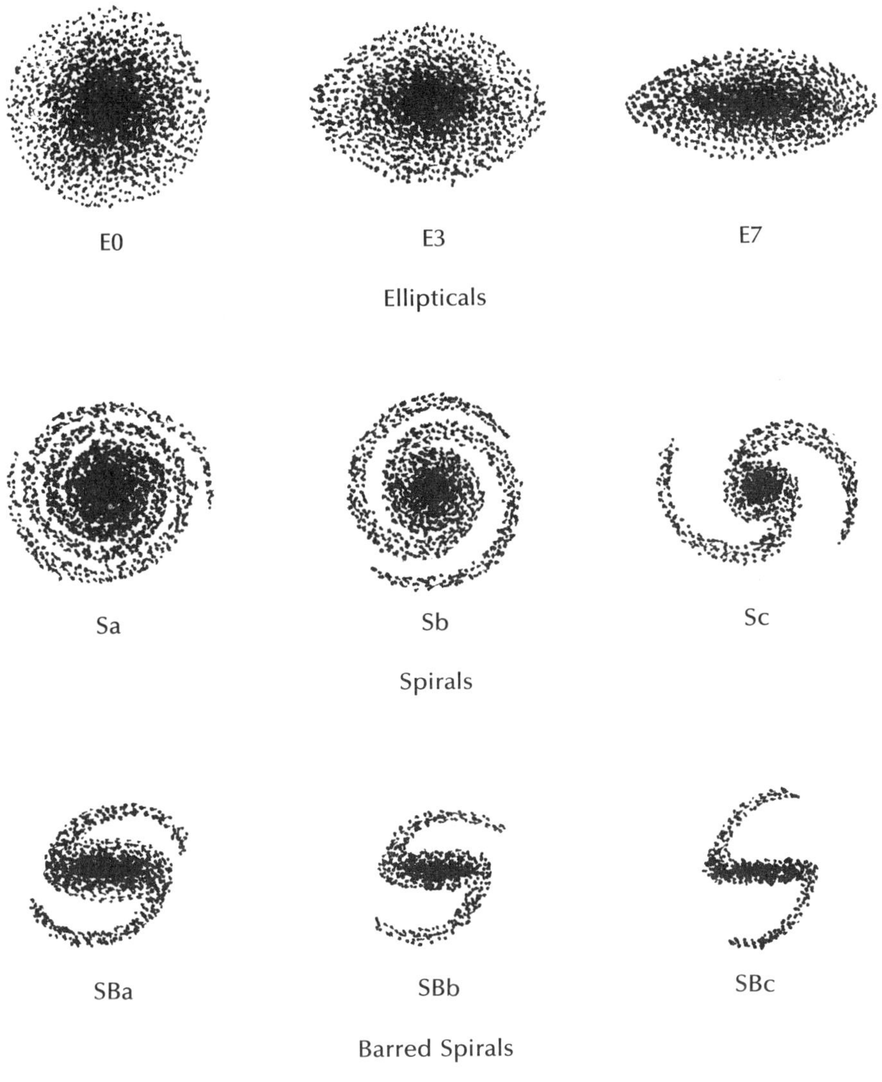

Figure 18
Classification of Galaxies

and the scale would go through 7, which would designate the longest and flattest elliptical shape. Thus an E0 would be spherical while an E3 would be moderately elongated and an E7 would be the longest elongation.

The ellipticals range in size from the giant to the dwarf size. The giant ellipticals are the most massive galaxies in the universe and are also very rare. They may contain up to 10^{13} stars. That's ten thousand billion stars. Their diameter may be one hundred thousand light years across, which is equivalent to the diameter of our own Milky Way galaxy. However, our galaxy has far less stars because it is a flat spiral.

There are a lot of dwarf galaxies, referred to as globular clusters. I would prefer to call them dwarf galaxies of the E0 type, since they are independent groups of stars. They may have 100,000 stars within them. They are the most common and probably amount to more than 50 percent of the total number of galaxies in the universe. All the ellipticals, when the larger ones are included, probably amount to 60 percent of the total. There are approximately thirty dwarf E0 galaxies (globular clusters) around our Milky Way. They are no doubt satellite galaxies revolving slowly around it and could have been captured on the Milky Way's journey through space.

The spiral galaxies are denoted by the capital letter S. They are further subdivided by the tightness of their spiral arms with the letters a, b, and c. A type Sa would be a tightly woven, high-spinning spiral while the Sc would have loose arms and be slow spinning. The spirals comprise about 20 percent of the total number of galaxies.

The barred spirals are classified by the capital letters SB, the S referring to "spiral" while the B refers to "barred." These also have a subclassification with the letters a, b, and c. If the central bar of the spiral is thick, it will have the letter a as the third letter of its classification. The c would designate a thin bar. Thus a thick-barred spiral would have an SBa classification while the thin-barred spiral would be SBc. These galaxies constitute about 10 percent of the total number of galaxies.

The irregular galaxies are just exactly what the word implies. They have no identifiable form. They generally appear as cloudlike in form. The two most commonly known are close to our Milky Way and are referred to as the Large Magellanic Cloud and the Small Magellanic Cloud. The irregulars are identified with the letters Irr.

HOW THE DIFFERENT GALAXIES ARE FORMED

The types of galaxies formed would depend mainly on four conditions. The first would be the shape and size of the cloud that the new galaxy is formed from. The second would be how the old galaxy penetrates the cloud or how closely it passes by. The third would be the type of old galaxy passing by. The fourth condition would be the relative velocity between the passing galaxy and the cloud.

An elliptical galaxy would be formed if the old galaxy moved through the center of the cloud. If the cloud was spherical in shape and the old passing galaxy was also spherical, then a new galaxy of the type E0 would be formed. If both the old galaxy and the cloud were gigantic in size, then a giant elliptical such as M86 or M87 in the constellation Virgo would be formed. (Refer to figure 19.)

To get a more detailed and better understanding of galaxies, the reader should refer to *The Hubble Atlas* by Allan Sandage. It provides beautiful pictures of various galaxies throughout the universe and their descriptions. Another book on galaxies is *Galaxies and Quasars* by William Kaufmann III. It contains a detailed explanation of their characteristics.

These giant ellipticals are generally in the center of clusters of galaxies, groups of galaxies in one region of the universe. They are in the center because they are the most massive and, as a result, the other galaxies revolve around them. The ellipticals appear to be completely devoid of gaseous matter—and they probably are. The reason for this is that the stars are rotating in opposite directions, the result of the central penetration of the cloud by the old galaxy, causing each portion of the cloud surrounding the center to rotate in opposite directions. Because of this counter-rotation, there would be some collisions among the stars causing ejections of gas and matter. However, because of the great distance between the stars, they would not be too frequent. This counter-rotation of the stars along with their planets would be responsible for sweeping the space clean of gases. The large stars could not condense any of these gases because of their high photonic pressure but the small, cool stars, along with most all the planetary bodies, could. The large planets within these galaxies would do a large amount of sweeping and condensing and could end up as small red stars. Since these galaxies contain a large number of red stars, that could be one of the explanations for their large presence. The fact that the ellipticals have a large number of these red stars does not mean that the galaxies are older, as the assumption would seem to indicate. The reason there are so many red stars is because more stars are formed as a result of all the gases being condensed and a lack of sufficient additional gas for them to mature to full-grown blue-white stars.

In the spirals, all the stars move in the same direction and the gases in between the spiral arms are carried along with the stars' gravitational pull. As a result the gases remain trapped and cannot be swept clean.

Figure 19
Giant Elliptical Galaxy in Virgo (NGC 4486, M87)
(© Kitt Peak National Observatory, Cerro Tololo Inter-American Observatory)

A lack of or a small number of blue-white stars in the spirals could indicate an older galaxy, but one in its middle stages rather than its latter stages of life.

As the old galaxy passes through the center of the cloud, it does not capture or condense any of the cloud's material because of the particle pressure as they are ejected from the neutron stars. This pressure would probably drive the material away from it. It would be after the old galaxy passes through the cloud that its gravitational pull would draw the center with it to start the rotation of the cloud. At the approach of the old galaxy, the cloud's material would be drawn toward it, but as the old galaxy passed through, the inertial resistance and reversal of direction would keep it from being dragged along with the old galaxy. It would, however, be pulled along the old galaxy's trail at a slower velocity as the galaxy was leaving. This would give the newly forming galaxy its initial movement.

If the cloud was small and an old dwarf galaxy went through the center of it, a dwarf globular galaxy would be formed. If a dwarf globular galaxy went through a large cloud, it would condense only a portion of it but would form a larger globular galaxy. There are probably a sizeable number of small or dwarf globular galaxies in intergalactic space that have not been captured as satellites of the larger galaxies.

If the cloud was elliptical in shape and the passing galaxy elliptical in shape or spiral in shape, a new galaxy of the type E1 to E7 could be formed depending on the length of the cloud relative to its width. Of course, either of the galaxies passing through the center of the cloud would also have a determination to its final shape.

If the cloud was spherical or elliptical in shape and the old passing galaxy elliptical in shape, and it went through the center of one half of the cloud, a new galaxy of the type known as NGC 5128 would be formed. This is in the constellation Centaurus. (Refer to figure 20.) The wide band surrounding this galaxy would indicate that the passing galaxy was elliptical rather than spiral.

If the cloud was spherical or elliptical in shape and the old passing galaxy spiral in shape, and it passed through the center of one half of the cloud, a new galaxy of the type Sa would be formed. A typical example of this type is NGC 4594 (M104) in the constellation of Virgo and it is called the Sombrero galaxy. (Refer to figure 21.) You will notice that this galaxy appears to be an E4 that is surrounded by a fast spinning dust band or spiral arms. The fact that the old galaxy went through a good portion of the cloud explains why the outer band appears to be and probably is spinning at a fast rate. Also the large nucleus indicates a deep penetration of at least one-half of the cloud.

Figure 20
Galaxy Type E0(pec) in Centaurus, NGC 5128
(© Kitt Peak National Observatory, Cerro Tololo Inter-American Observatory)

Figure 21
Galaxy Type Sa/Sb in Virgo (NGC 4594, M104)
(© AURA Inc., Kitt Peak National Observatory)

Notice how smooth and thin the band appears. This would indicate that the passing galaxy was spiral rather than an elliptical.

A spiral galaxy of the type Sb would be formed if the old galaxy was a spiral and passed just outside the cloud. An example of an Sb is NGC 4565 in the constellation Coma Berenices. (See figure 22.)

NGC 598 (M33), known as the Triangulum galaxy in the constellation Triangulum, has a very slow spin and probably is not very flat. This galaxy is classified as an Sc. (Refer to figure 23.) This type of galaxy was formed by an old galaxy that passed at a considerable distance and therefore has not imparted too much of a spin within it.

The relative flatness of a newly created galaxy and its velocity of rotation would depend on how close the old galaxy passed by. Also, its passing velocity would be a factor. An old galaxy passing by at close range would also generate a greater number of bright stars in the newly formed galaxy. Generally, the brightest side or arm of the new galaxy would indicate on which side it was bypassed, the brightest side or arm being on the same side as the passing galaxy at the time of its formation.

FORMATION OF BARRED SPIRALS

Barred spirals obviously would be formed from elongated clouds that would look like long streaks in space if they could be seen. How did these clouds come about in the first place?

I believe that these clouds formed from the trails left by the old galaxies as they moved through space ejecting the neutrons, protons, and electrons. These streaks of particles would form new gaseous atoms and molecules, as I've mentioned previously. Then they would acquire additional matter and dust that was moving slowly through space. Eventually the trails would be disturbed and broken up by the various gravitational fields from the different galaxies passing by at various distances. Of course, some portions of these elongated clouds would remain intact for some time and be the clouds principally involved in the formation of the barred spirals, as an old galaxy passed by in the distance. These elongated clouds are probably the youngest clouds in space. The spherical clouds would be the oldest because they had sufficient time to evolve into their shape by remaining intact for a long time. The barred spirals seem to be smaller than the regular spirals and that could be expected if they were formed from the younger clouds.

The thickness of the cloud would probably determine whether it would result in an SBa, SBb, or an SBc.

Figure 22
Galaxy Type Sb in Coma Berenices, NGC 4565
(© Kitt Peak National Observatory, Cerro Tololo Inter-American Observatory)

Figure 23
Galaxy Type Sc in Triangulum (NGC 598, M33)
(© Kitt Peak National Observatory, Cerro Tololo Inter-American Observatory)

The old galaxies must have bypassed these clouds at the point where the arms are bent because it left the central bars relatively undisturbed.

The galaxy NGC 7741 in the constellation Pegasus is a typical example of a barred spiral. It is an SB type. (Refer to figure 24.)

These streaks of clouds seem to be fairly common because of the number of barred spirals. They also seem to be part of some of the regular spirals, because a large number of these regular spirals have but two arms tightly wrapped in the regular pattern.

Figure 24
Barred Spiral Galaxy in Pegasus, NGC 7741
(© Palomar Observatory)

Chapter 5

The Big Bang

It seems to me that the big bang is a totally illogical concept. As I've mentioned in chapter 4, matter does not explode spontaneously. Collisions seem to be the cause of all explosions in space. Matter itself builds up and disintegrates slowly. It also does not congregate in one central region due to the inertia from the motion of the bodies. This would seem to contradict the big bang theory that the universe started from a condensed singularity. In this instance, the word "singularity" meaning that there was one condensed congregation of mass in the entire universe before the explosion.

Matter will congregate or merge if the bodies are moving in the same direction and are sufficiently close for their gravitational fields to draw them together. This only happens, though, under certain conditions, as I've explained previously in chapter 4, during the galaxy formations.

Gases, on the other hand, have a tendency to disperse. They do not condense by themselves. Gas molecules behave like electrons. Electrons are negatively charged particles and they repel each other. Gases have the same characteristics. They have a tendency to occupy all space because of this. Solid bodies the size of the Moon or Mercury are not capable of containing any gases to their surface. And if these large bodies cannot contain any gases, how can gases compress into a singularity by themselves? That obviously seems to be another impossibility. However, in intergalactic space, where the temperature is close to absolute zero, smaller bodies could attract gas molecules.

In nature the gases can compress into stars only if there is a massive solid core that would be much larger than the earth because under those conditions it would be in a heated state. The other characteristics of star formation would also have to be involved as I've explained previously for the star to build up.

Astronomers today say that the universe is the same regardless of which direction we look or how far we look. This is known as the *cosmological principle.* Of course, we have to except the presence of the band of stars that constitute the Milky Way galaxy. The cosmological principle pertains to only what is beyond our Milky Way. Even beyond our Milky Way there are some slight variations in the distribution of matter but not to any great extent. We can then say that the universe is homogeneous and isotropic, homogeneous meaning that matter is distributed equally no matter how far we look and isotropic meaning that matter is distributed equally no matter in which direction we look.

This type of universe then complies more with the steady state theory than with the big bang theory. The way matter would be distributed in the big bang theory will be explained later on.

THE RED SHIFT

The red shift is the primary reason that led scientists to the concept of the big bang. But before I go any further, I would like to explain to the reader what the red shift is.

The most common example of the explanation of the red shift is the approaching train with its horn blowing. As the train approaches the listener the sound is a little higher in pitch, and when it passes the sound suddenly drops in pitch. The reason for this is that the sound-waves are compressed during the train's approach and are being stretched out as it is moving away. An illustration of this is in figure 25.

The wavelengths of light are similarly affected by the movement of the emitting source. When a star is moving away from the earth, the star's spectrum moves toward the red portion because its light waves are being stretched out. When it is approaching, the spectrum moves toward the blue because the light waves are being compressed.

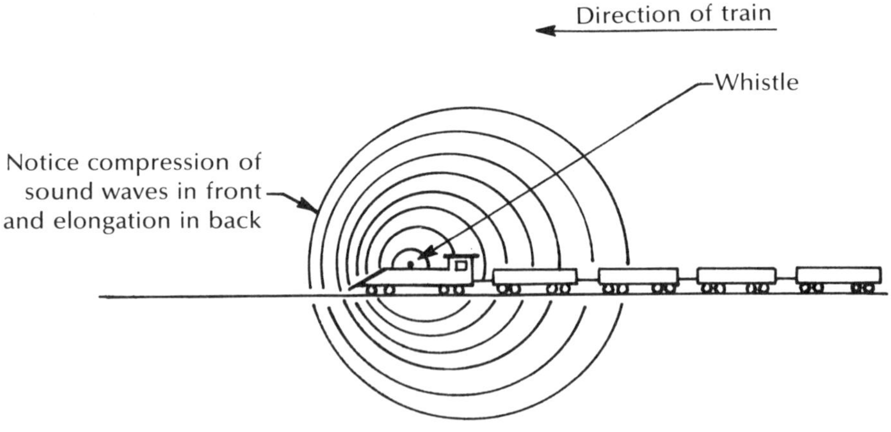

Figure 25
Basic Principle of Red Shift

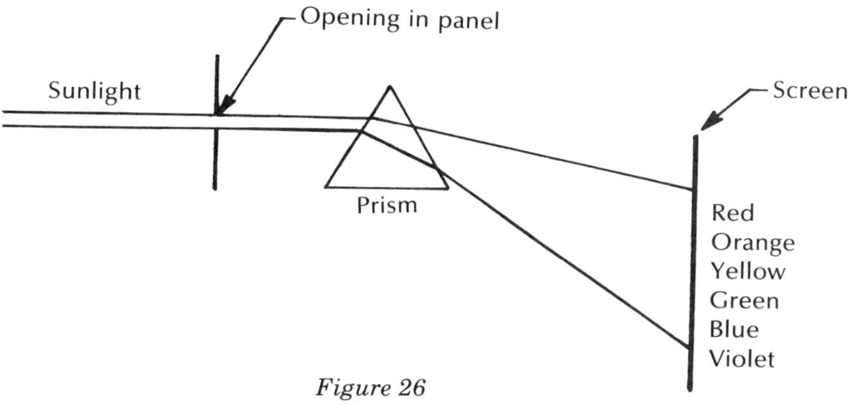

Figure 26
Visible Light Spectrum

Now, to give a more detailed explanation of the red shift, I'll have to start with the sun's light spectrum. When a person is seeing a rainbow on a rainy day, while the sun is shining through a clearing, that rainbow is a separation of the white light into the various color components that comprise it.

This can also be done in the laboratory by the use of a prism, as illustrated in figure 26. This white light from the sun is separated into the colors red, orange, yellow, green, blue, and violet and is known as the continuous light spectrum, and also as the visible light spectrum. The red light has the longest wavelength and the lowest frequency, while the violet has the shortest wavelength and the highest frequency. The other colors vary in wavelength progressively according to their position in the spectrum. The visible light spectrum is just a small part of the entire electromagnetic spectrum.

Another characteristic of the sun's spectrum is the presence of dark lines at irregular intervals. These are known as absorption lines and denote the presence of various elements. These lines are specific wavelengths of light absorbed by the various elements in the sun's lower atmosphere. Each element has its own identifying spectrum just as each human being has his own characteristic fingerprints.

Each element also emits light in the same wave length as it absorbs, when it is heated to a point where it can emit light. These lines are known as emission lines and are bright, while the absorption lines are dark.

It is really the position of these emission and absorption lines rather than the colors themselves that tell if there has been a shift in the spectrum of the body being observed. The lines most commonly used for this purpose are the H and K lines of singly ionized calcium, which are the most prominent lines in the sun's spectrum and therefore the most easily identifiable. These two lines tell us how far they moved and in which direction, although the lines of other elements can also be used. By measuring the amount of the shift, the assumption is that the velocity of nearby stars and distant galaxies can then be determined. However, this can be done only for objects that are approaching or receding. It cannot be used for objects that are moving crosswise.

The shift in the lines is determined by taking a photograph of the spectrum or whatever light source one wants to analyze. This photograph is known as a spectrograph. If a comparison is then made to the sun's spectrograph, a shift in the lines can be detected. If the lines are shifted toward the red end of the color spectrum, then the object is considered to be red-shifted and is receding from the earth. If the lines are shifted toward the blue end of the spectrum, then the object is considered to be blue-shifted and is approaching the earth.

Since practically all the galaxies in our universe are red-shifted, this would imply that the universe is expanding since all the galaxies would be receding from the earth. This single set of observations, then, is what initiated the big bang theory. One notable exception is the great galaxy in the constellation of Andromeda. This galaxy's light is blue-shifted and is supposed to be approaching our galaxy at about 260 km per second.

The red shift, I believe, can be interpreted in another, more realistic manner. Since the most distant objects are assumed to be receding at the highest velocities because of their large red shifts, it seems to make more sense to conclude that if an object is farther away, the red shift could be the result of the light slowing down rather than the object's velocity away from the earth, the slowing down being due to the gravitational pull of the object. Galaxies have a tremendous amount of mass. The combined gravitational force of all the stars within that galaxy is constantly effecting its pull on those particles of light heading away from it. Gravitational force extends to infinity, so it is constantly slowing down the light that leaves it and therefore creating the illusion that the galaxy is receding. The truth probably is that the distant galaxy is moving at the same velocity as, say, M87 in the constellation of Virgo. Therefore, my contention is that the red shift is predominantly caused by the gravitational force of the galaxy that the light has left and is not due to its receding velocity from the earth.

A very good example to support my contention is the galaxy NGC 7603. It is composed of two galaxies very close to each other, with a thick, curved chain of stars connecting the two. (Refer to figure 27.) It

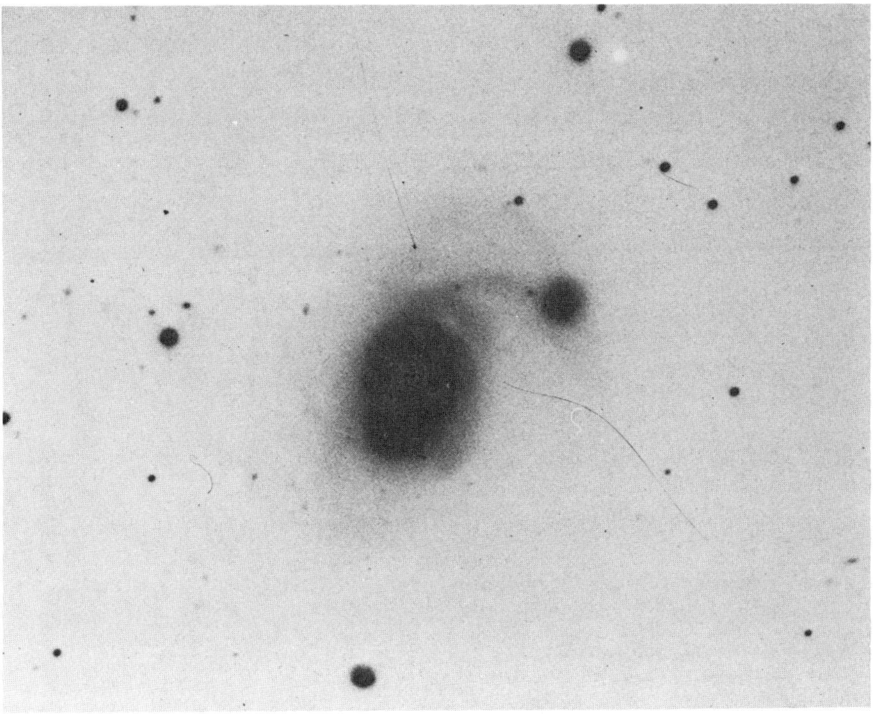

(Negative image)

Figure 27
Galaxy, NGC 7603 (© Dr. Halton C. Arp)

is plainly obvious that the two galaxies are interrelated and at the same distance. And yet the large galaxy has a red shift that would indicate a receding velocity of 8,700 km/s, while the small companion has a red shift that indicates a receding velocity of 16,000 km/s. Since the larger galaxy appears to be a flat spiral viewed from its axial point, then the density and depth of the stars would be rather shallow and a lower red shift would be the result. The smaller companion, however, appears to be an elliptical galaxy. This galaxy would have a greater density and depth of stars in relation to our line of sight. Therefore it would create a greater red shift than the spiral. This then creates the illusion that the smaller galaxy is receding at a greater velocity.

There are other such situations where there is no doubt that two galaxies are linked together by a bridge of stars indicating that they obviously are at the same distance but at the same time have different red shifts.

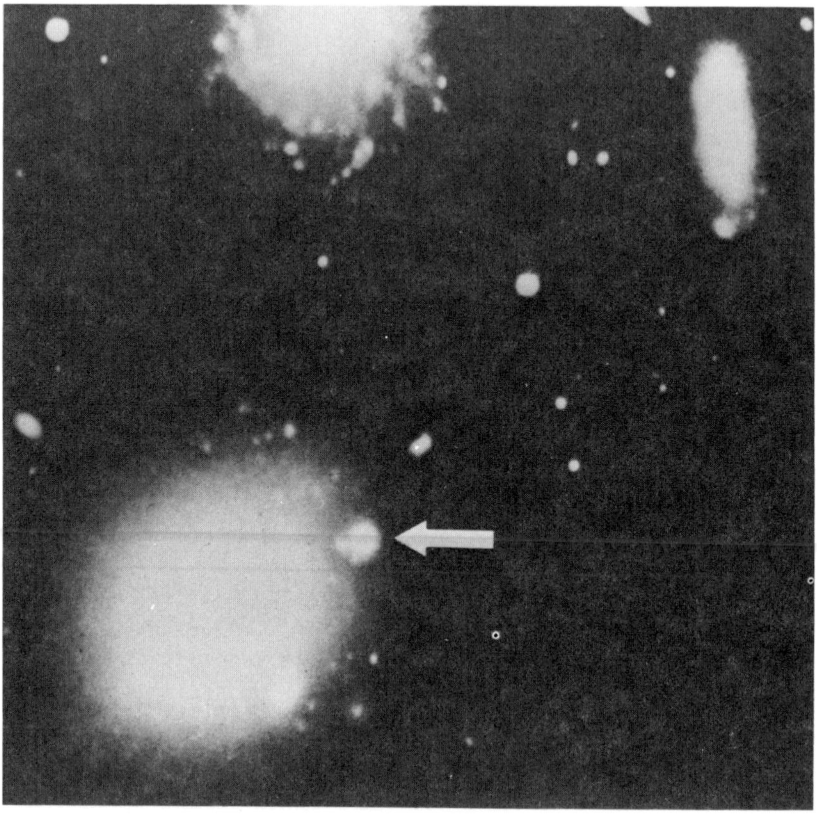

Figure 28
The arrow points to the quasarlike object that appears to be in front of the large galaxy (NGC 1199). Notice the shadowy separation between the two. (Courtesy of Dr. Halton C. Arp)

An example of a different nature is the galaxy NGC 1199. A small quasarlike object to the right appears to be in front of this galaxy. You'll notice a slight shadowy arc separating the two that could be caused by the light of NGC 1199 being refracted by the quasar as it passes by. The quasar has a red shift of .044, while the galaxy has a red shift of .009, which is considerably less even though it appears to be behind the quasar. The red shifts indicate that NGC 1199 should be closer rather than further away. (Refer to figure 28.)

We should reasonably assume that all smaller galaxies are revolving in elliptical orbits around the most massive galaxy within any particular cluster, the central and most massive galaxy moving in a much smaller orbit within the center of the cluster. Then we can say that all small clusters of galaxies are revolving around the largest cluster of galaxies within their own region of the universe, while the largest cluster of galaxies is revolving around the most massive concentration of galaxies in the universe, which would be the center of the universe.

Disregarding the red shift and the expanding universe, then, we can say that half the galaxies are approaching us and half are receding from us, probably at the more sensible velocities of 500 to 1000 km/s. The approach or recession would depend on which part of its orbit each galaxy was in.

The red shift then should be used to establish distances of galaxies but not velocities. Also, since the red shift would be different for elliptical galaxies in relation to spiral galaxies viewed from their axial point, then that would have to be taken into account. A spiral galaxy viewed from its equatorial plane could possibly have a red shift that is close to or equal to the elliptical's red shift.

THE BLUE SHIFT

The Andromeda galaxy and our Milky Way galaxy are part of a small cluster of galaxies known as the Local Group. The Andromeda is two million light years from the Milky Way. This is a very small distance when you consider the astronomical distances between the various galaxies within our universe. This is also the most massive galaxy within our Local Group. It is 160,000 light years across compared to the Milky Way's diameter of 100,000 light years. I would estimate that it is more than four times as massive as the Milky Way. That would mean that the Andromeda is the central galaxy within the Local Group. (Refer to figure 29.)

The light spectrum from the Andromeda is blue-shifted and therefore is supposed to be approaching our Milky Way at about 240 km/s. It would be more appropriate to say that our galaxy is approaching Andromeda at a velocity that's greater than that galaxy is approaching ours. Why aren't these two galaxies separating in accordance with the big bang concept that all galaxies are receding from each other? The obvious answer is that these two galaxies are locked in elliptical orbits around each other by their respective gravitational fields. The other galaxies in the Local Group are also no doubt revolving around Andromeda. The galaxies within the Local Group are not separating, I'm sure.

The Andromeda and Milky Way galaxies are probably on the return cycle of their orbits, and that would account for their approach toward

Figure 29
Great Galaxy in Andromeda (NGC 224, M31)
(© Palomar Observatory)

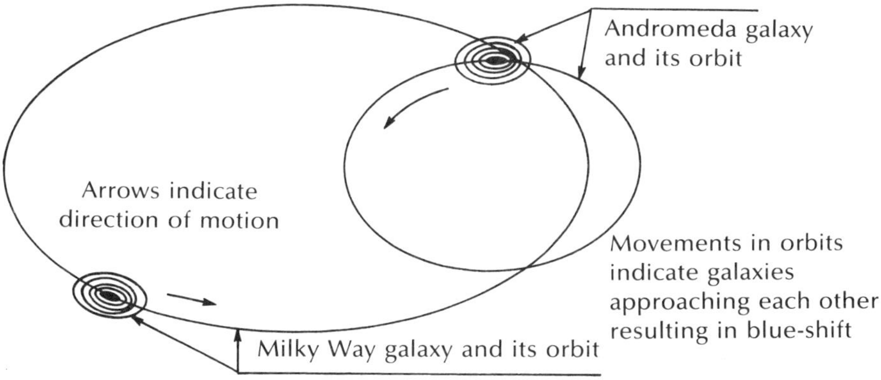

Figure 30
Probable orbits of Andromeda and the Milky Way disregarding all other galaxies within the Local Group

each other. (Refer to figure 30.) After they pass their perihelions within their respective orbits at a certain point they will begin to recede from each other and go to the outer fringes of their orbits at their aphelions. However, they will again approach each other after they pass their aphelions. Their combined orbital velocities would have to be greater than the 240 km/s that is now calculated directly between the two. Of course, their velocities in orbit would vary depending on where they were with the highest velocity at the perihelions and the slowest at the aphelions.

Another factor, but of much less significance, would be the shift of Andromeda's light toward the blue by the gravitational force of the Milky Way. This however would constitute a very small fraction of the total shift.

OLBERS PARADOX

Olbers paradox is a concept that is used to substantiate the idea of the expanding universe. However, this can be explained easily.

The concept of the paradox is that the night sky should be as bright as the daytime sky, this being so because of the even distribution of the stars in the sky no matter where we look or how far we look.

The inverse square law states that the intensity of light will be reduced to one quarter of its strength at twice the distance and one ninth of its strength at three times the distance, and so forth.

However, since the number of stars will increase by four at twice the distance and by nine at three times the distance, the light reaching us should be equal at all points no matter where we look or how far we look. So the entire sky should be bright.

Since this is not so, the assumption is that the expanding universe is responsible for the dark sky. Because the stars and galaxies are receding from us, the energy we receive from them is proportionately reduced.

First of all, the vast majority of the visible stars in the sky are not receding from us because they are contained in our own galaxy. Some of these stars are actually approaching us while some are receding, but at very slow velocity. It depends on which direction you look in the sky. Yet we know that we cannot see all the stars in our own galaxy.

The primary reason for this is that there is a lot of dust and gaseous molecules in our surrounding space to obscure these distant stars. The principle is the same as driving a vehicle in a fog. One can see a certain distance in a fog even though the fog is evenly distributed around us. However, the light from the greater distances has to pass through a greater number of water particles and therefore is obscured. So what we see at the greater distances is the water particles themselves.

One argument against this hypothesis is that the dust and the gas molecules continuously absorb energy from the stars and therefore should eventually heat up and glow with sufficient intensity to be visible. However, I'm inclined to think that those particles transform the visible light to infrared or use that energy to form atoms, molecules and more complex molecules. It's simply the process of reconstituting matter.

PRIMORDIAL BACKGROUND RADIATION

A recent discovery to support the big bang theory is that a radiation from the big bang explosion still exists out there in space. This is known as the background radiation and is considered to be diffused throughout the entire universe. Its temperature is around 3K and its wavelength about 7.5 cm.

The 3K means a temperature of 3° above absolute zero. This is the scale that is used in astronomy. It can be easily converted to the Celsius scale by simply adding 273. Zero on the Celsius scale is equal to 273 on the Kelvin scale, which is the freezing point of water. Zero on the Kelvin scale, of course, is equal to absolute zero, which is considered to be the lowest temperature that can be obtained. The Fahrenheit scale we need not be concerned with here because it is not used in any scientific work. So the Kelvin degree is equal to the Celsius degree except for their numerical differences. An illustration of the comparison between the three scales is in figure 31.

This 3K background radiation temperature has supposedly cooled out to its present temperature from the big bang's temperature of billions of degrees. The expansion of the universe is what caused the cooling out. The first second after the explosion, the temperature was estimated to be around ten billion degrees K. The present 3K background temperature was determined from some measurements and observations. This background radiation has been used to establish our galaxy's velocity through absolute space at about 600 km/s. It was done by measuring variations in the 7.5 cm wavelength in different directions around our earth.

I believe, however, that this temperature exists in the immediate vicinity of the space surrounding our solar system and is not necessarily uniform throughout the universe. The interstellar space temperature in the center of our galaxy or close to the center would surely be higher than it is around our solar system. On the other hand, the intergalactic space temperature in our universe is in all probability very close to absolute zero, possibly 1/2K, since it is completely devoid of stars.

The 3K background radiation has a wavelength of approximately 7.5 cm. This means that the radiation from the big bang, which was

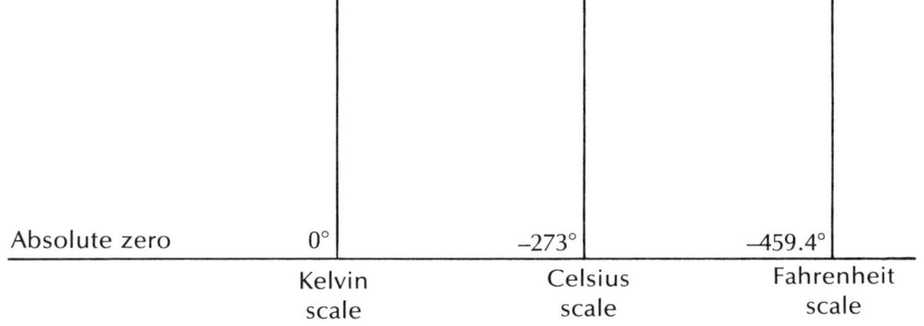

Figure 31
Comparison of Temperature Scales

in the billions of degrees and obviously of very short wavelength, such as gamma rays, X rays, and light waves, has changed frequency to the present longer wavelength. The wavelengths of gamma rays, X rays, and light waves is from one one-hundred-millionth (1/100,000,000) of a centimeter to sixty six-hundreds of one hundred-millionths (6600/100,000,000) of a centimeter long. That means that the big bang radiation has increased its length as it cooled out from 750 million to 120 thousand times its original length to its present length of 7.5 centimeters. That could mean that all electromagnetic radiation increases in length as it leaves its source and expands into outer space. Therefore, the wavelength of light does not remain constant, which could mean that the red shift would be caused by the expansion and the increased length of the light waves, this being added to the slowing velocity resulting from the gravitational pull of its source.

Another thing I would like to point out is that the background radiation left over from the big bang should not be in our part of the universe but at the outer periphery of the present universe. Since it would not be slowed down by gravity as much as the galaxies, it would have to be a considerable distance beyond our galaxy in relation to the center or origin of the explosion. (Refer to figure 36, p. 81.)

THE ISOTROPIC NATURE

The universe seems to be isotropic throughout. This means that there is matter scattered throughout the universe in equal proportions regardless of which direction we look or how far we look. We know that there are concentrations of galaxies known as clusters in some regions of the sky, but when we take the totality of the universe into account we can be sure that clusters of galaxies are scattered evenly throughout. Does this even distribution of matter support the big bang theory? I'm inclined to think it does not. A big bang universe would have to be in the shape of a hollow sphere. It cannot have any other form. All the matter would have to be receding from the center of the explosion.

Current astronomical reasoning among the proponents of the big bang says that there is no center to the explosion, that it was not like the explosion of a bomb, that the entire universe was filled with infinite matter and energy; that the big bang was the creation of space itself; that in the beginning, matter was compressed to infinite density while taking up all space. It has no center or edge. It was an explosion of space itself.

Those ideas cannot be realistic because I'm sure no one can comprehend the meaning of those ideas. Space cannot be compressed. How can one compress a dimension like one kilometer to a dimension such as one centimeter? The meaning of the dimension is then changed, not compressed. Dimensions cannot be changed, otherwise they would have no literal meaning. In the beginning, space was infinite in its dimensions and infinite it shall remain.

How can anyone imagine matter condensed to infinite density and then have it occupy infinite space? I'm inclined to think that is an impossibility. If matter is condensed to infinite density, then that matter must occupy less space. That must be a simple law of physics, otherwise the *density* would have no meaning.

Neutron stars are composed of matter compressed to almost maximum density that can be had and yet they still have dimensions even though they occupy practically the minimum space allowable. If you shrink both matter and space together, then the density would remain the same and you could not start with infinite density. The origin of the big bang is inexplicable in rational terms.

Now, if we come to the more reasonable conclusion that the big bang did start out as a typical explosion from one central region of the universe, then we would have to be receding from that center at a certain velocity. That velocity cannot be ascertained because we do not know the total amount of mass at present in the universe. Not knowing how much mass there is, we cannot determine how much effect the gravitational force had in slowing down the dispersal of the matter. From the very beginning after the explosion, the matter was receding at the velocity of light. Now, if we use the present velocity of 600 km/s, as established by the background radiation variation, as our recession velocity from the center, then we can assume that we'll reach zero velocity in the near future. (Refer to figure 32.)

According to the Hubble constant, though, the galaxies are receding from each other at 55 km/s per megaparsec. It is based on the observations of the galaxies' red shifts. This constant may have been recently revised downward to 45 km/s per megaparsec.

Since *mega* is a prefix denoting one million, and parsec equals 3.08×10^{13} km or 3.26 light years, then a megaparsec would be equal to one million parsecs, or 3.08×10^{19} km.

All the galaxies are supposed to be receding from each other at that velocity equally in all directions and it would be the same regardless of where you were in the universe. This means that from any point in the universe, matter recedes from that point equally in all directions. Obviously, then, there is no center to the universe or all points would

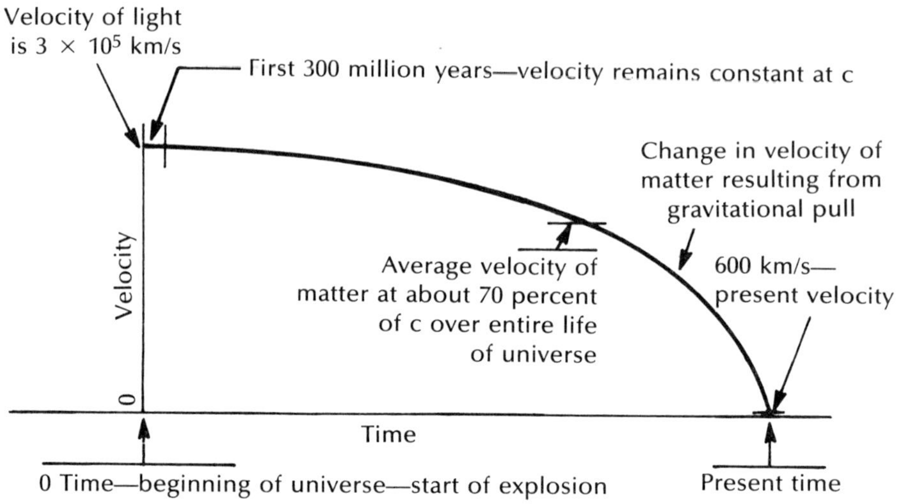

Figure 32
Velocity Curve of Matter in the Universe from Beginning to Present

be central. This concept also says that the farther an object is away from any point, the faster it is receding. This just conveniently tries to circumvent the idea of a central point in the universe.

To place ourselves in the center of the universe is an extremely remote possibility. Aristotle placed the earth in the center and had every other body in the heavens circling it. This false idea dominated scientific thinking for 1,800 years. Of course, everything seems to be revolving around the earth because the earth is spinning on its axis, creating the illusion that we are the central body. Aristotle's mistake was that he depended on observational evidence only. Incidentally, Aristotle was one of the greatest Greek philosophers of his time who lived around 350 B.C.

It wasn't until the sixteenth century, when Nicolaus Copernicus presented his heliocentric theory explaining that all planets revolved around the sun, that seemed to clear up a lot of problems created by the Aristotlelian theory. It was after he published his book that astronomy made great strides in the study of the heavens. Copernicus, of course, was the great Polish astronomer who lived around A.D. 1500.

Now let us return to the concept that the big bang happened like a nuclear explosion. During the first few seconds, days, and even years after the explosion, the particles of energy would have to be moving away from the center of the explosion at the velocity of light since they were all photons, electrons, and neutrinos, according to the big bang physicists. Shortly thereafter, neutrons and protons are formed. Regardless of what particles there were, the important thing to establish is the initial radial velocity that these particles had. And that velocity was at the speed of light, which is designated by the letter c.

Next, how long does the propulsive effect of the explosion exert on these particles before gravity begins to take over and they begin to slow down? We'd have to assume that these particles formed gases as the temperature began to cool down. Also, the gases would continue to expand as there was available space toward the outer perimeter of the newly forming universe. Gases do not gravitate toward each other except possibly very weakly in intergalactic space, where the temperatures are very close to absolute zero.

As the temperature continues to cool, heavier elements would have to start forming from the gases before gravity would have any retarding effect. The question is, when did the heavy elements begin to form or whether they would even form from the gases alone? In the steady state theory, we don't have this problem, since all the elements existed throughout eternity.

The duration of the explosion would determine how thick the shell of this exploding universe would be at the start. Further expansion of the shell would continue as long as the temperature was sufficiently hot for the gases to continue expanding.

Let's assume that the explosion lasted for one hundred years. (Refer to figure 33.) At that moment in time, the temperature would have dropped to about 20,000K. Then the further expansion of the gases because of the high temperature would continue for the next 150 million years till the temperature dropped to 300K. This is slightly higher than room temperature. At this temperature, I presume the gases would still continue to expand slowly until the temperature dropped to 100K. At this point the gases would cease to expand because heavier elements would start to form and gravity would start to take effect. This would be 300 million years after the explosion. The thickness of the shell then would continue to expand even after gravity had taken over because the matter still has momentum. The perimeter of the universe would likewise continue to expand due to the momentum. However, at this stage the matter is in free flight as gravity begins to slow down the expansion. The expansion of the shell's thickness, 300 million years after the explosion, is probably less than one percent because the residual temperature and momentum of the gases is very small. 300 million years is a very small part of the total life of the universe which is estimated to be between 12 billion and 20 billion years old. (Refer to figure 34.)

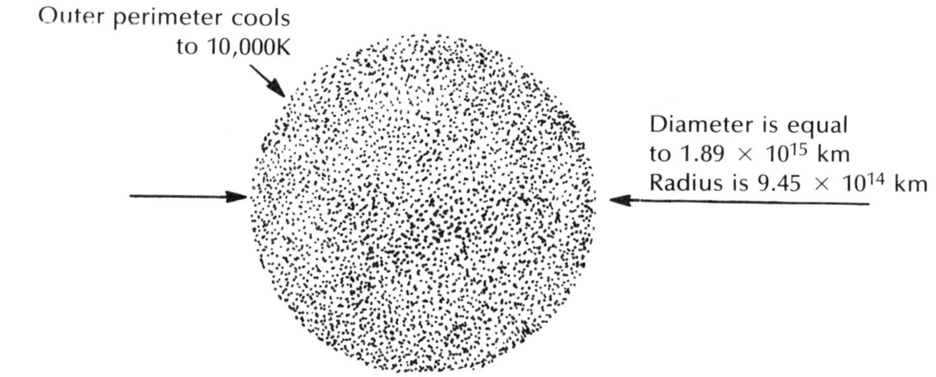

Figure 33
Big Bang Universe during First 100 Years
Cross-section of universe during first 100-year duration of the explosion. Notice that matter occupies entire space.

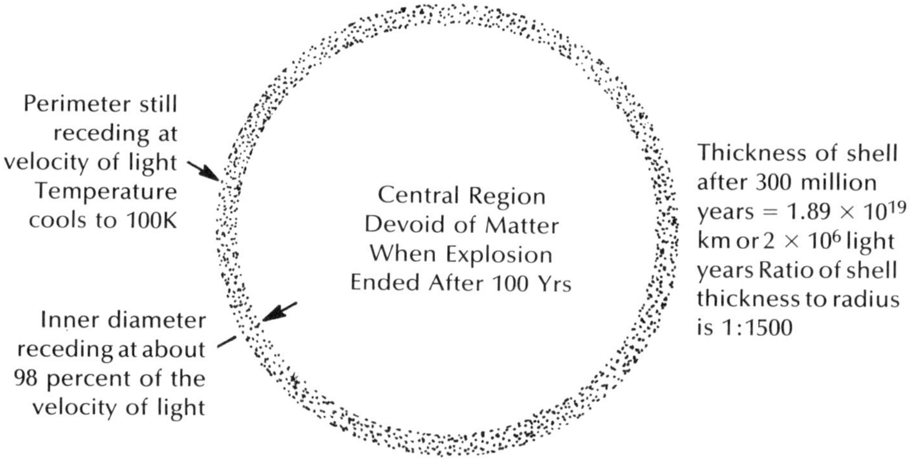

Figure 34
Big Bang Universe after First 300 Million Years
After 300 million years, periphery expands to 5.67×10^{21} km, or three million times larger than the original diameter after 100 years. Radius = 2.84×10^{21} km.

For 300 million years, then, the particles and gases were receding from the center at the velocity of light. Thereafter, the velocity began to recede slowly for the next 11.7 billion to 19.7 billion years, as gravity began to take effect. Now, if our present velocity is 600 km/s, as I've previously mentioned, and we do reach zero velocity in the near future, then the galaxies will start to converge back toward the center again.

Up to the present time, then, how do we establish the age of the universe? It's pretty obvious that the only observation we have to do that is the red shift and the subsequent Hubble Constant that resulted from it. If we were to assume that the galaxies are receding from each other at the periphery of our present universe according to that constant and at the same time disregard the gravitational pull, then it would be easy to establish the age of the universe. All we have to do is divide one megaparsec by the 55 km expansion rate that occurs every second and we get an age of 5.61×10^{17} seconds, or 1.779×10^{10} years: 17.79 billion years. (Refer to figure 35.)

Now if we go back to figure 32 and from the nature of that curve estimate the average velocity of this matter from the beginning to the present at about 70 percent of the speed of light, then our big bang universe should have a radius of 1.178×10^{23} km (70 percent of speed of light $\times$ life of universe in seconds). This would be the point where our earth is from the center of the explosion and would constitute the present size of our universe with the present calculated velocity of 600 km/s relative to absolute space. You'll notice in figure 32 that the velocity of the matter does not reduce uniformally in a straight line but rather in a curve similar to a parabola, because of the increasing gravitational pull of the matter as it formed. This is why I've used the 70 percent figure. The equivalent size of this universe in light years is 1.246×10^{10}. (Refer to figure 36.)

The greatest actual dispersal velocity between the galaxies would be along the shell and travelling along the axial lines spread the furthest apart from each other, because they would be moving in different directions and that would mean all galaxies on the opposite side of the explosion. These galaxies, however, would not be seen because their light has not been able to reach us at the present time. The galaxies above or below our own and on the same or nearby axial lines would have zero or the slowest dispersal velocities, since these galaxies would be moving in the same directions.

With this kind of universe, then, we would not be able to see any quasars, which are the most distant galaxies—beyond, say, fourteen billion light years distant—because from two to three billion light years must have been traversed by the particles before any galaxies or qua-

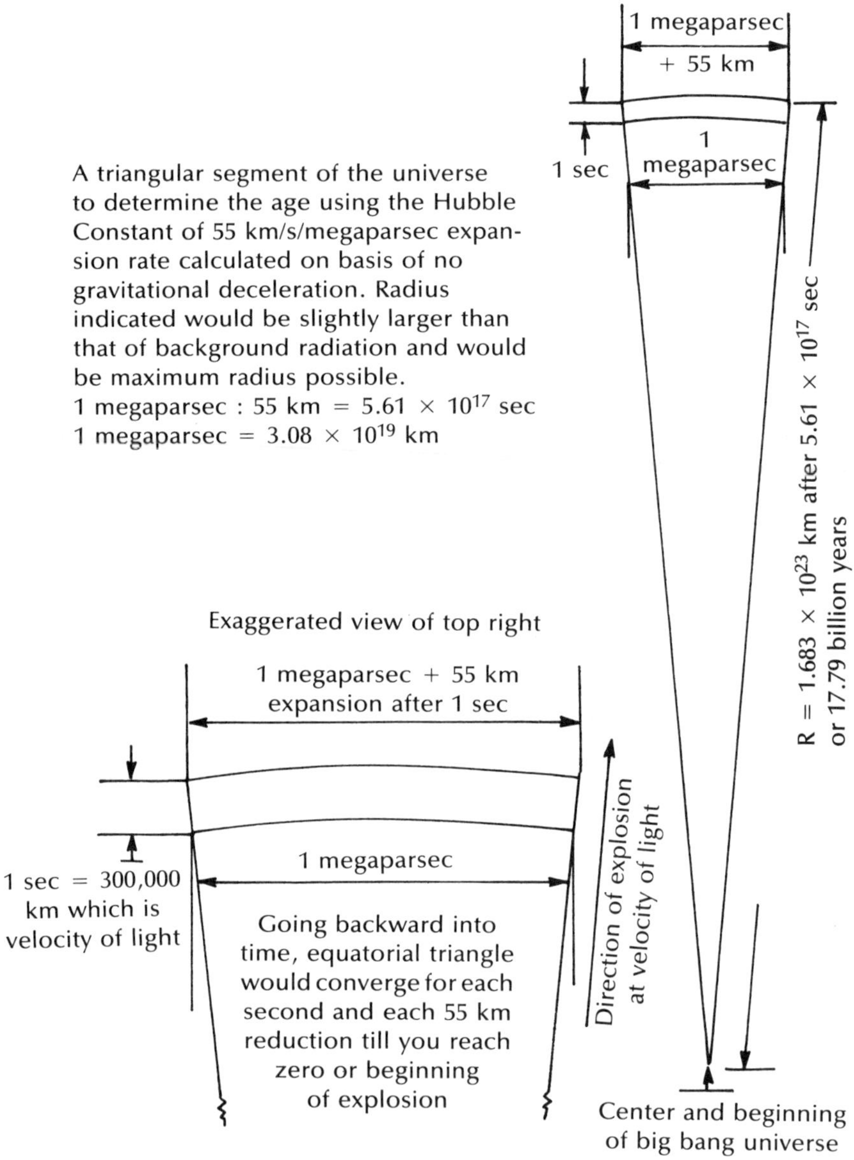

Figure 35
Determining the Age of Big Bang Universe

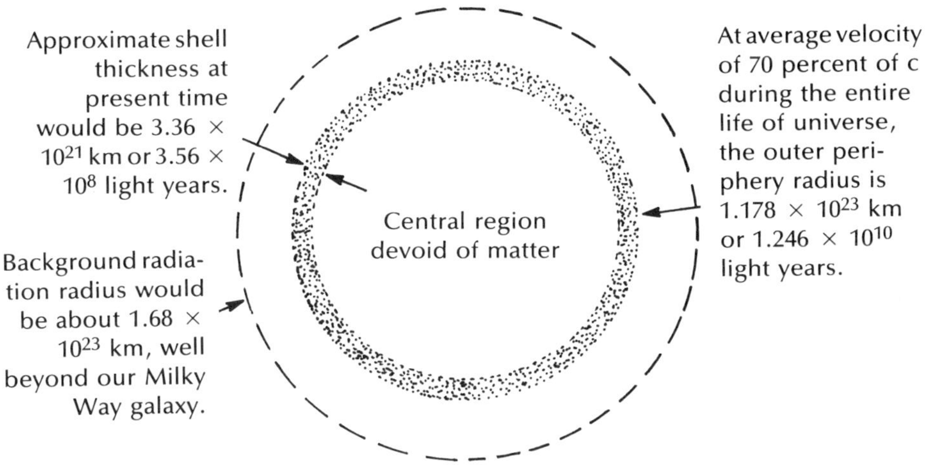

Figure 36
Dimensions of Big Bang Universe at Present Time

sers formed. We should also see these galaxies in only one region of the sky. This region would be toward the center and around the region of the explosion. Also, the galaxies would be approaching or receding from us at an angle or would appear to be standing still if they were moving in a direction perpendicular to us, since most are close to doing so, such as 3C 48, 3C 147 and 3C 123. 3C 9, which is supposed to be 14 billion light years distant because of its large red shift and subsequent high velocity of 80 percent that of light, should have a velocity of about 20 percent that of light because its angular recession from us would not be very great. (Refer to figure 37.) The distant galaxies closest to us should actually be slightly blue-shifted, such as 3C 48 and 3C 147. Notice that the present universe is a spherical shell while the visible universe is shaped somewhat like a pear.

To get back to the observed isotropic nature of the universe, we see quasars at distances of up to 16 billion light years, such as OH 471 with its very large red shift. Other quasars are closer but scattered in different regions of the sky and all have red shifts. So these are examples of the impossibility of the big bang universe.

To sum it all up, the two greatest pieces of observational evidence to support the big bang universe is the red shift and the 3K, 7.5 cm background radiation.

As I've said before, the red shift is due to the distance of the galaxies and their gravitational force acting upon the light. This would mean

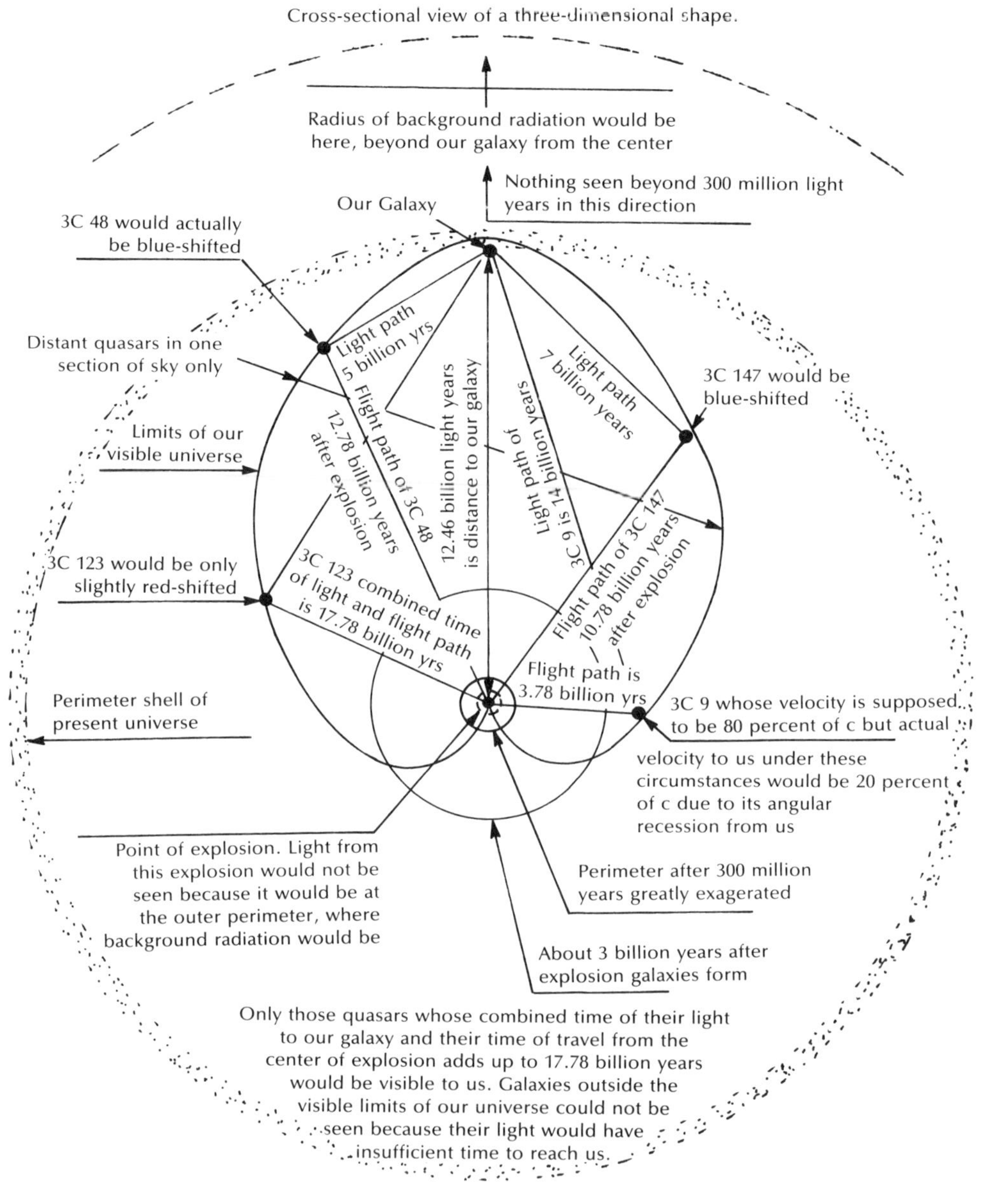

Figure 37
Present Shape of Visible Universe according to Big Bang Theory

that the velocity of the light relative to our earth is the cause of the red shift and not the recessional velocity of the galaxy. The 3K temperature is due to the space temperature in our area of the Milky Way galaxy rather than a uniform space temperature throughout the universe left over from the big bang. Another important factor is that the background radiation, if it ever existed, would be on the outer perimeter of our universe and not scattered among the present galaxies. The outer perimeter of this background radiation would have a diameter of around or close to the maximum diameter of 3.36×10^{23} km because it would be the least affected by the gravitational forces. That would place it at about 4.8×10^{22} km beyond our own galaxy at the extreme outer perimeter of our universe. (Refer to figure 36.)

Chapter 6

Black Holes

Black holes are supposed to be large collapsed neutron stars that have such strong gravitational fields that even light cannot escape. The theory is that when a giant star goes through the supernova stage (nuclear explosion) and blows away all the remaining gas that surrounded it, the rapid cooling of the remaining neutron core causes the collapse of the mass and the core condenses to a singularity. The central portion of regular stars is supposed to be at enormously high temperatures that keep the central core from collapsing during the regular life of the star. A singularity is a tiny speck of mass left in the center and is supposed to have infinite density. This density is so enormous that the singularity is practically crushed out of existence and an event horizon is formed around it.

The event horizon is a sphere surrounding the singularity that is the upper limit that would constitute the light's escape altitude before it falls back to the singularity again. In other words, the light has sufficient velocity to escape to a certain height before the gravitational force of the singularity overpowers the light and causes it to fall back again. Some particles of light would reach the uppermost altitude of the event horizon and circle the singularity indefinitely locked in continuous orbit at that altitude. (See figure 38.)

Since a black hole is supposed to acquire mass and never lose any, I would wonder why we never have any massive black holes that can be seen. Even though a black hole would not emit any light, its tremendous gravitational force would bend any light passing close to its event horizon and so would appear to be a dimly lit star from a great distance. From closer observation, it would appear to be a halo and the size and brightness of the halo would depend on the size of the singularity's mass and the number and size of the stars in its vicinity. (Refer to figure 39.)

Of course, these halos would have much smaller diameters than regular stars, but if any were nearby they would be observed. Also, if a spectrograph was taken of such a star, it would have a mixture of light from the surrounding background of stars. This would give it a unique characteristic relative to the other regular spectrums of ordi-

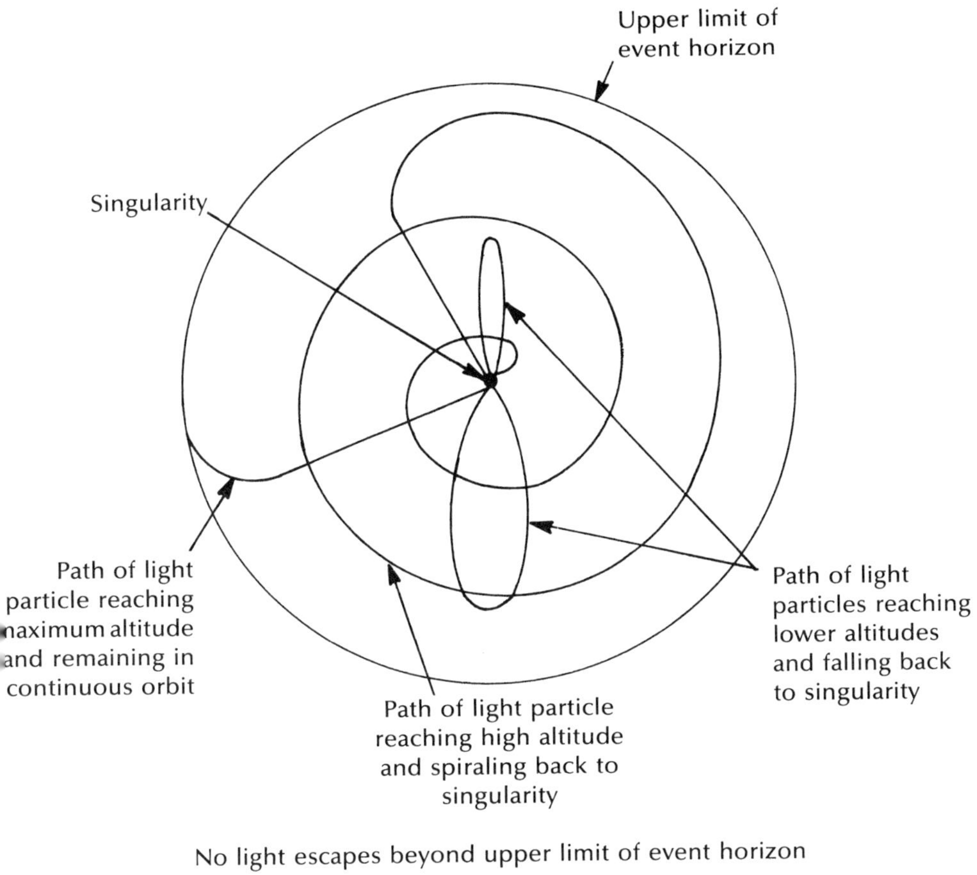

Figure 38
Black Hole

nary stars. Stars would also mysteriously shift their positions as a black hole passed their line of sight relative to us.

There is also the contention that black holes occupy the nucleus of galaxies. If this were true then the obvious conclusion would be that the black holes would pull all the stars in the central region toward themselves. Under these circumstances, the black holes would grow to such enormous sizes that the nuclei of these galaxies would be dark instead of light because of the enormous pull of the gravitational field upon the light.

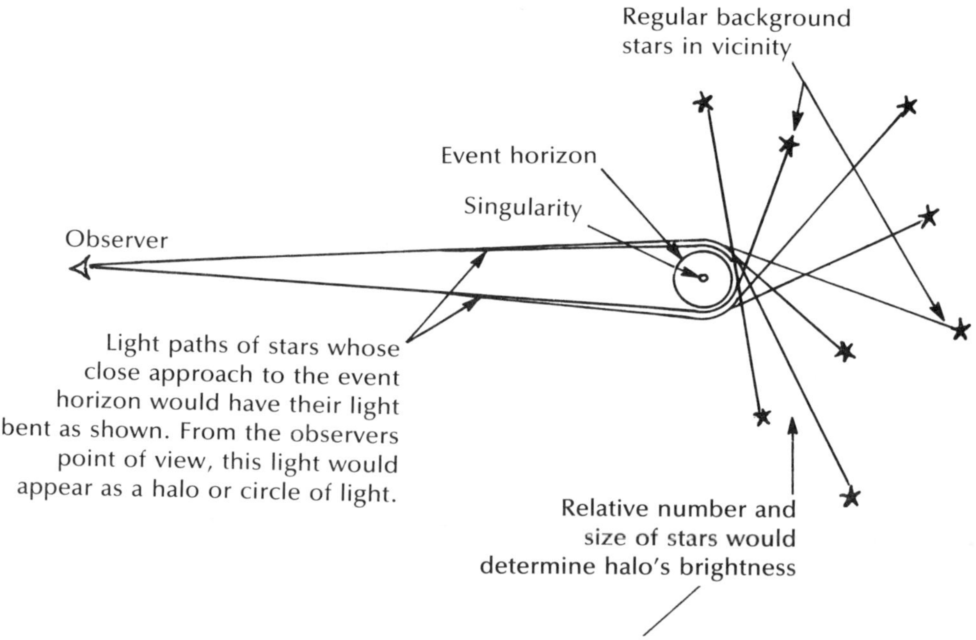

Figure 39
Possible Visibility of Black Hole by Formation of Halo

However, the nuclei of galaxies are extremely bright. This would mean that there is a high concentration of stars burning at a much faster rate than normal because of the high space temperature in the centers. And, no doubt, high concentrations of massive stars would exist because of the great possibility that they merge there more frequently. Because they would be enormous in size, they would burn out more rapidly, and as I've said before, they would disintegrate completely. We can then conclude that the very bright nuclei of galaxies are the result of high concentrations of massive stars which contribute to high space temperatures in those regions resulting in their rapid disintegration into light, energy, and particle matter.

Another likely place to look for neutron stars is at the outer perimeters of spiral galaxies, because those are the regions where the largest number of type O and B stars exist. Because they are the largest and would obviously burn out the quickest, these stars would be the likeliest candidates for the formation of black holes. Their residues or hulks should be more common in those regions. However, since stars burn out completely, I'm inclined to think that black holes are nonexistent in those regions as well.

Since black holes would continue to grow with time to gigantic masses and enormous sizes, they would have been observed by this time. Since none have been observed, then they cannot exist.

Spontaneous nuclear explosions that lead to the formation of black holes do not occur in nature. Supernovas are the result of collisions resulting in a chemical explosion from the mixture of the different gases because of that collision. Nuclear explosions are the work of humanity, and because of this I'm afraid we've modeled our universe after them.

The centers of stars are not at the enormously high temperatures that we are led to believe. After a star reaches its maximum size during a galaxy's formation and no further condensation takes place, the high temperature in the central region that develops from the buildup ceases to rise and the central temperature then begins to gradually cool down over a period of time by conduction and convection to the outer surface. (Refer to figure 40.) The center may then reach a temperature of about one or two thousand degrees higher than the surface.

I'm sure everyone is familiar with the way heat is conducted through metals. One portion of the metal can be heated and the heat is soon distributed throughout the entire piece. This is known as *conduction.* The same principle would occur in the solid central core of the stars because that material is considerably more dense than any metal. This conduction then would radiate the heat to the liquid portion of the star.

Heat within liquified gases, such as water, rises to the top and stays there. So if heat is applied to the bottom of a vessel containing any liquid, the heated portion of the liquid rises to the top while the colder portion settles to the bottom. This then causes the liquid to circulate until it reaches a uniform temperature throughout. This is known as *convection.*

The heat from the upper surface of the star's core would be carried to the surface of the star though convection. This heat would then be contained at the surface till it reached a high enough temperature for light to be radiated, and the surface of the star would then lose mass by the radiation of that light into space.

COLLAPSE OF CENTRAL CORE

Now to get back to the collapse of the central core. There are two stable primary particles of matter known as the proton and the electron. The proton weighs 1.6726×10^{-24}grams and is approximately 2×10^{-13} centimeters in diameter. That gives it a density of 4×10^{14} grams per cubic centimeter. The electron weighs 9.1096×10^{-28} grams

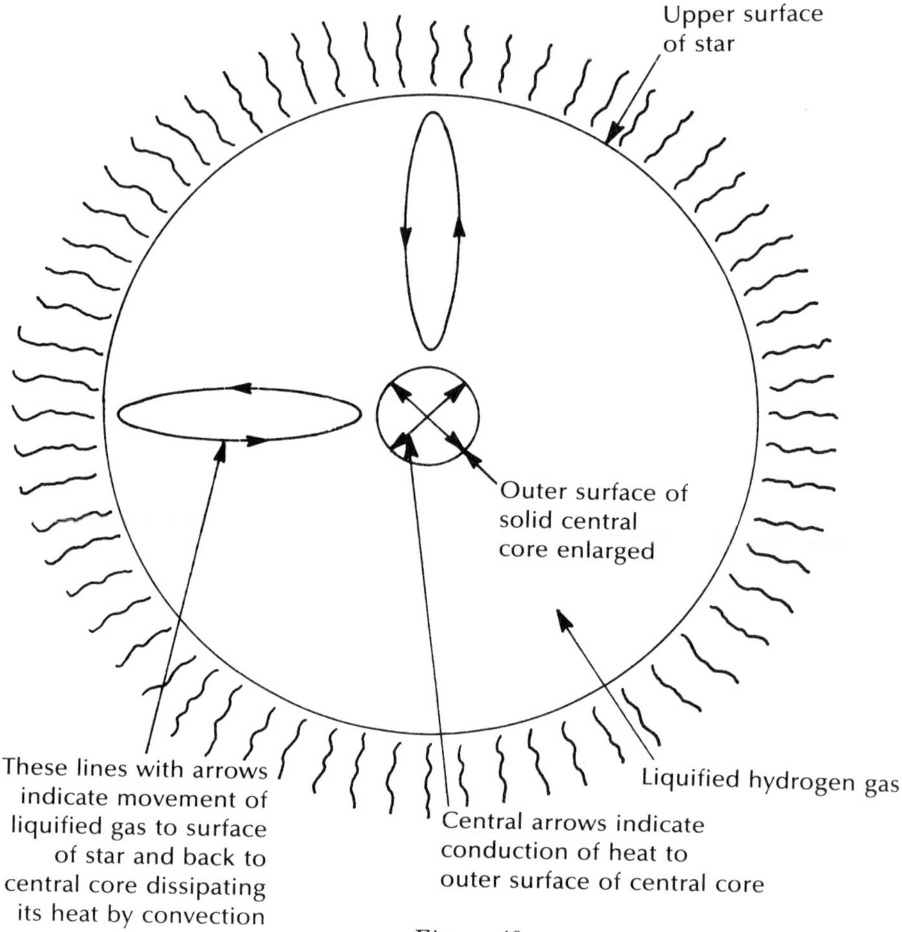

Figure 40
Distribution of Heat within a Star

and is approximately the same diameter as the proton which is 2×10^{-13} centimeters. That gives the electron a density of 2.17×10^{11} grams per cubic centimeter. The proton is 1,836 times heavier than the electron.

To give the average reader an idea what all this means, we'll compare those figures to cast iron, which has a density of 7.2 grams per cubic centimeter. That means that the proton is 5.55×10^{13} times heavier than iron. That is 55.5 trillion times heavier. The reader should know that one trillion has twelve zeros after the one. The electron mass is 3×10^{10} times heavier than iron. That is 30 billion times heavier.

With such enormous densities, how can one expect those particles of matter to collapse?

Sea water at its greatest depth supports one million times its own weight and it is hardly heated from the great pressure. The temperature of sea water at its greatest depth is just 4° centigrade, which is just 4° above the freezing point of ordinary fresh water. It's obvious that the heat generated at those great depths is a lot less than the heat dissipated at the surface by radiation. Since the water does not seem to collapse from the great pressure, how can one expect the primary particles like the proton and electron to collapse?

Water has a lot of empty space between its primary particles since water weighs only one gram per cubic centimeter. That means that water is nothing but pure space between the protons and electrons, and yet it can withstand the pressure of one million times its own weight without collapsing.

I'm sure iron can withstand much more times its own weight than water can without collapsing, since its density is 7.2 times greater. These two substances are not solid matter, like the proton or electron, because there is a lot of empty space between their particles. The proton and electron, we'll have to assume, are solid substances with no empty space within them. If there is no empty space within them then they cannot collapse.

Black holes are supposed to form from neutron stars of three or more solar masses. This would give them a diameter of about 40 km for an 8 solar mass star. The central cubic centimeter of mass in such a star would then be supporting 2 million times its own weight on both sides for a total of 4 million. How can anyone expect this neutron mass within the center of such a star to be crushed by weight that is only 4 times greater than the mass that water can withstand? It is conceivable that these extremely dense particles of matter can support not only millions but billions of times their own weight.

The cores within the stars are condensed to enormous densities and the close proximity of the atoms would destroy a lot of electrons as the result of collisions. The particles from these collisions would no doubt become photons and would radiate to the surface as heat. The electrons would be vulnerable to collisions because of their high velocities as they circle the protons. The protons, on the other hand, would not be so vulnerable because they would have very little motion compared to the electrons since their masses are much greater. That would mean that the neutron stars have a lot of free protons scattered throughout the central mass. These free protons would also keep the cores of these stars from collapsing due to the electrostatic repulsion between them. Then, when the weight and material above the central core has been radiated and dissipated into space as energy, this repulsion would actually contribute to the disintegration of the central neutron core till nothing is left of it. The protons and neutrons expelled from a disintegrating core would go on to form new atoms.

Chapter 7

The Wave Nature of Particles of Matter

The tiniest particles that compose matter travel in straight lines when in free flight. That is, they appear to travel in straight lines but they move in a wave pattern in their line of flight. This chapter is an attempt to explain this wave phenomena, and is hypothetical and does not necessarily represent fact.

First of all, I would like to explain to the reader the basic fundamentals of the atom and electricity.

MATTER

All the matter that we see and feel around us is composed of atoms and molecules. Molecules are two or more atoms coupled together by electrostatic forces.

The atom is the smallest individual composite unit of matter or substance known. With the exception of the hydrogen atom, which has one proton and electron, all the other atoms are composed of protons, electrons, and additional neutrons. The protons and electrons are the smallest stable particles in existence. The neutron is stable only when coupled with a proton within the nucleus, which is the central region of the atom. It is also electrically neutral while the proton and electron are electrically charged. Outside the nucleus, the neutron's life span is only about fifteen minutes. The electron circles the nucleus at an orbit that may vary in size depending on the energy or, more specifically, the velocity that it is moving at.

The simplest atom known is the hydrogen atom. Other atoms, such as those of the heavier metals, can be composed of hundreds of individual particles.

The proton carries the positive charge and the electron carries the negative charge. It is these electrical charges that hold the electron in orbit around the proton because they are opposite charges and attract each other. Similar charges repel each other. The electron revolves around the proton because it is a much lighter particle. Although both particles are physically the same size, which is 2×10^{-13} cm in diameter, the proton is 1,836 times heavier. 2×10^{-13} cm is equal to a decimal point followed by 12 zeros and then a 2. That means that if these particles were stacked side by side, it would take 5,000,000,000,000 to equal one centimeter. There are 2.54 cm to the inch. We are dealing here with extremely tiny particles of matter.

These particles however are extremely dense. If we were able to pack as many protons as we could into one cubic centimeter of space, these protons would weigh 2.4×10^{14} grams. That's 2.4 followed by thirteen zeros. To make it more understandable, they would weigh 4.3 billion tons per cubic inch. If we were to do the same with the electrons, they would weigh 2.36 million tons per cubic inch. This gives you an idea of the enormous density of these particles.

The electrostatic force that holds the atom together is still more gigantic. We all understand the force of gravity since we weigh ourselves every now and then, and most of us are familiar with the moon being held in orbit around the earth and the earth in turn being held in orbit around the sun. This electrostatic force is 7.3×10^{36} times stronger than the force of gravity. That's 73 followed by 35 zeros.

The diameter of the hydrogen atom in its lowest energy state is 10^{-8} cm. That's when the electron is moving at its smallest orbit. If we were to compare the size of the atom to our solar system, and made both the proton and the electron the size of the sun, the atom would be six times larger than the outermost fringe or diameter of our solar system. This, of course, would be the orbit of the planet Pluto. In between these two particles would be nothing but empty space. So you can see that even matter has nothing but empty space between the particles that compose it. Hydrogen gas of course is the lightest of all the elements.

THE ELECTROMAGNETIC FORCE

Everyone is familiar with static electricity. I'm sure everyone who has attended a high-school physics class has seen experiments conducted with it.

Lightning during a thunderstorm is a typical example. It is a discharge of static electricity between positively and negatively charged regions. If there is an abundance of electrons in one region and a deficiency in another region, whether in the air or on the ground, then those electrons will be strongly attracted by the deficient region. The result is a bolt of lightning. The thunder is the resultant sound from the discharge.

Static electricity may play a part in the gravitational force between the various planetary bodies and stars in our universe. It also plays a dominant part in the formation of stars, as I've explained in chapter 4.

The magnetic force is the other part of the electromagnetic force. It is the magnetic portion that plays a strong part in the wave characteristics of electrons, photons, and other particles. It is generated by the movement of the particles.

Magnetic force has two polarities rather than the one that electrostatic force has. These polarities are known as the north and south poles. Magnetic force always has two polarities no matter how many times you divide the substance. Even if reduced to the tiniest amount, that amount will still have two polarities.

Magnetic force has been a great boon to humanity because it is used in the generators that produce our electricity, and the motors that supply us with our power to make our work a lot easier. It is used in practically all the electronic components that we have and use in our lives.

The nature of electromagnetism is the movement of electrons through wires or space and the resultant magnetic force fields that result from the motion. It is the movement of electrons that generate the force fields, or for that matter any charged particles. However, we'll concern ourselves only with the electrons. When the electrons are motionless, no magnetic field exists.

Electric circuits all have a source of voltage that drives the electrons within that circuit. And the flow of the electrons is always from the negative terminal of the voltage source to the positive terminal. There are some minor exceptions in physics and transistor technology, but we need not be concerned with that here.

When electricity flows through a circuit, a magnetic field builds up as illustrated in figures 41 and 41a. You'll notice that if you were to imagine placing your right hand around the wire with your thumb pointing in the direction of the current flow, the fingers will indicate the direction of the magnetic lines of force. The magnetic lines of force are a part of the magnetic force field that surrounds the wire. This is known as the right-hand rule. Though you cannot label these lines of force as being of a specific polarity, such as north or south, the fact that they do have direction is an indication that they do have polarity.

A permanent bar or horseshoe magnet also has lines of force between the poles as illustrated in figures 42 and 42a. You'll notice that the movement of the lines of force is from the north to the south pole. If we place a wire carrying an electric current in the magnetic field of the horseshoe magnet, the wire will move depending on the direction

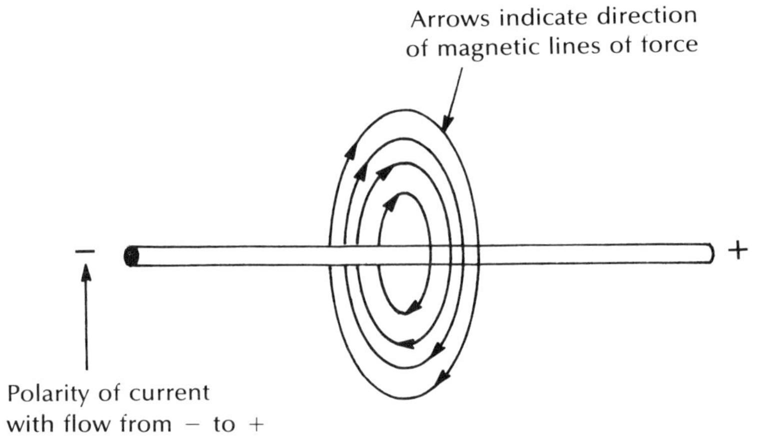

Figure 41
Magnetic Field around Wire Angular View

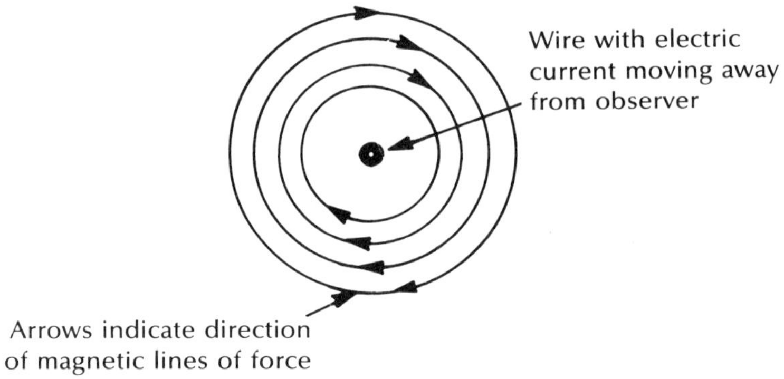

Figure 41a
Magnetic Field around Wire, In-line View

of the current and the direction of the magnetic field as illustrated in figure 43. You'll notice where the lines of force generated by the wire and the lines of force generated by the magnet cancel out, the wire will be forced in that direction. Where the lines reinforce each other, the magnetic lines force the wire away from it. This is the principle of the electric motor.

The principle of the generator is more or less just the opposite. In this case, mechanical power is applied to the rotor of the generator and the wires within the rotor moving through a field generate electricity instead.

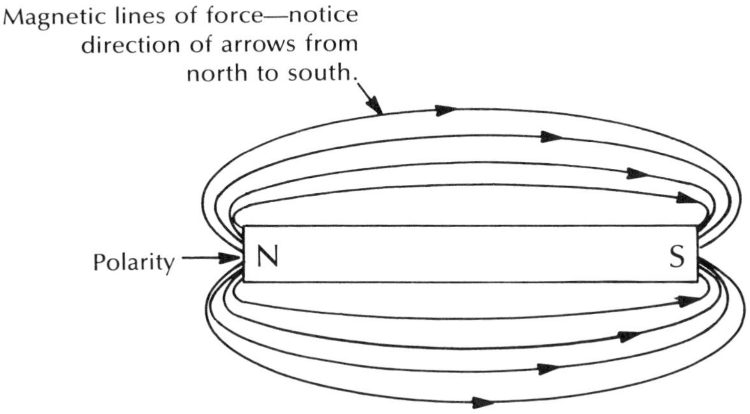

Figure 42
Magnetic Field around Bar Magnet

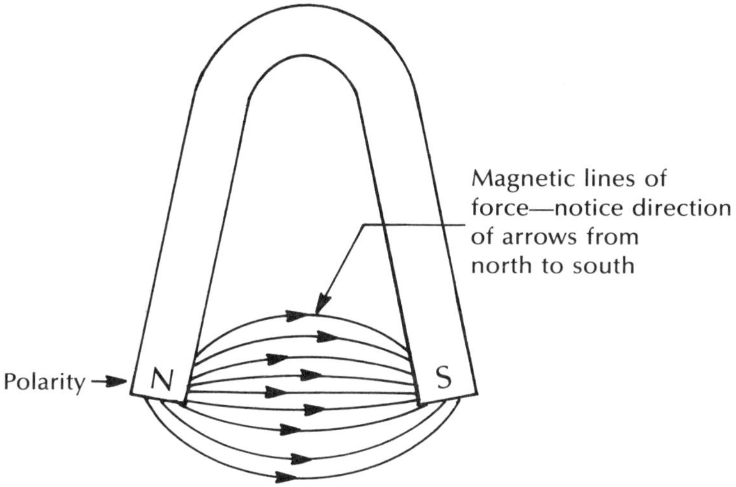

Figure 42a
Magnetic Field of Horseshoe Magnet

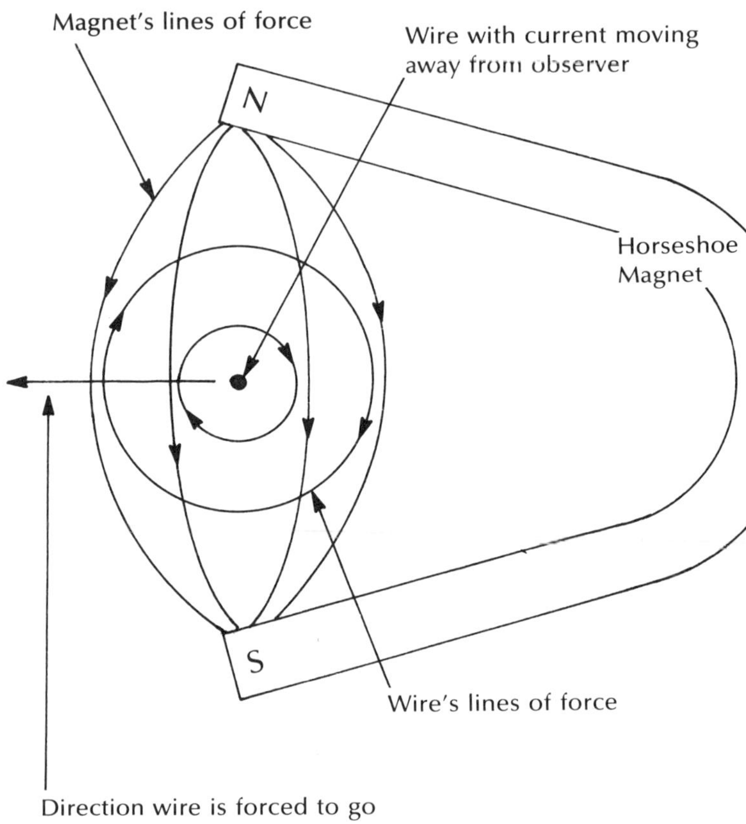

Notice that on right-hand side of wire magnetic lines reinforce each other, while on left-hand side they oppose each other and therefore cancel out, forcing wire in that direction.

Figure 43
Principle of Electric Motor

PHOTONS OR ELECTRONS THROUGH SPACE

The same rules can apply to the movement of photons or electrons through space. The photon is a negative particle just like the electron since I believe it is a tiny fragment of the electron. Since it is an electrically charge particle, it will affect any other photon nearby because of the respective magnetic fields generated by their movement.

Let's assume two photons are moving away from their source of light and are practically parallel to each other. There will then be a magnetic field surrounding each one. Since they are moving in the same direction, the lines of force around each will also be moving in the same direction. (Refer to figure 44.) The resulting lines of force will then cause the photons to move toward each other, because the lines of force between them will be cancelling each other out and the subsequent central field will be greatly reduced or nonexistent. The outer lines of force then move the photons together as illustrated in figure 44a. As the photons are forced together they continue to approach each other until the electrostatic force is sufficiently strong enough to repulse them, since they are both negatively charged particles and like charges repel each other. Their momentum, however, will continue to move them toward each other for a short distance until the electric force field is strong enough to reverse their direction. (Refer to figure 44b.) The electric force field will act like a spring. As the direction of the particles is reversed and they move away from each other, the electric force field will then be losing its effect on them and the magnetic force field will stop their separation and will reverse their direction again. The magnetic force field will also act like a spring as it bounces the particles together again as illustrated in figure 45.

This would then explain the particle wave nature of matter. The wave frequency would depend on the velocity of the particles and their subsequent magnetic field strength as well as their mass and subsequent electrostatic field strength.

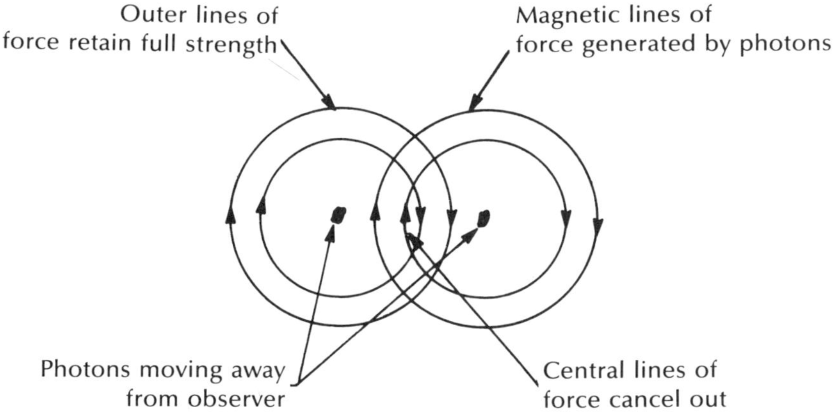

Figure 44
Oscillating Photons in Space

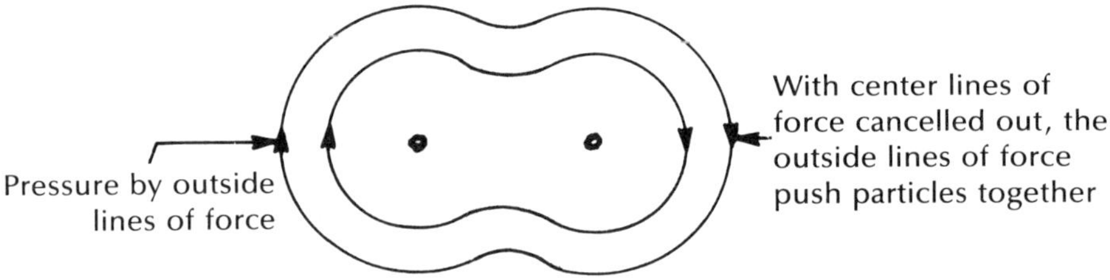

Figure 44a

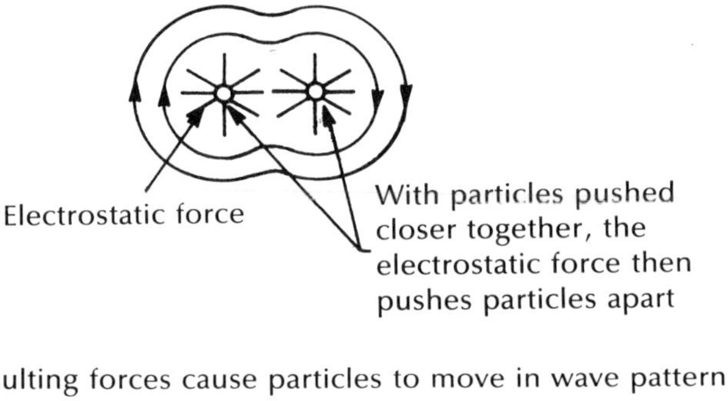

Figure 44b

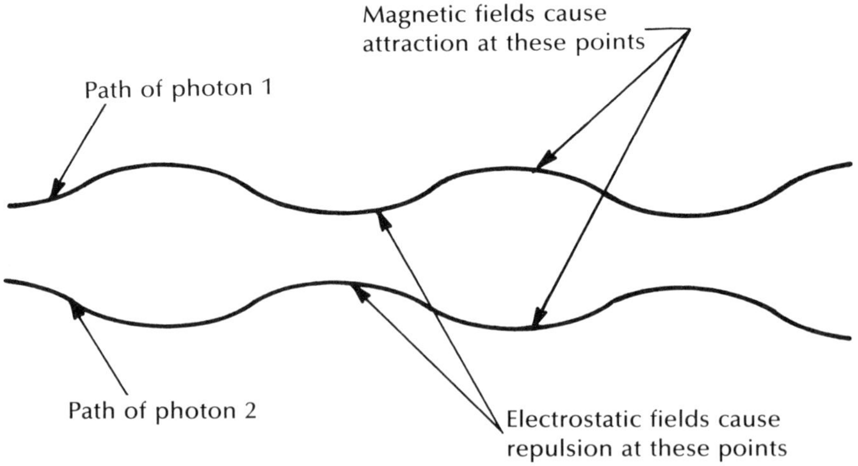

Figure 45
Wave Created by Photon

The same action would take place with any two similarly charged particles whether photons, electrons, or positrons.

Two oppositely charged particles would also cause a wave. If an electron and a positron were moving along side of each other, the different force fields would then do just the opposite by reversing the actions. The attracting force would be the electric charge and the repelling force would be the magnetic field. However, these two particles could end up revolving around each other and eventually merge to form a neutral particle. This happens only when they are moving around a common center, in other words when they have approximately equal masses. The strong attraction by the electric charge would be reinforced by the magnetic field since it would also cause a strong attraction. The two opposite electric charges would cause magnetic fields of opposite polarity and so the magnetic fields in the axial center would be cancelled out causing the particles to be pushed in toward the center. This combined strong force could cause the particles to revolve at extemely high velocity before they did merge. This would happen as I have said only when the particles are in orbit rather than moving parallel to each other. Like particles would never wind up in orbit around each other because of the repulsion of the electric charges, and in this instance becauses the magnetic fields would also cause a repulsion.

The wavelike action of these particles in space can be compared to the oscillations of electrons in radio or television transmitters' oscillating circuitry. This principle here is not exactly the same, but the principles are similar because of the electromagnetic forces acting in unison.

A simple oscillating circuit is a coil of wire known as an induction coil and a condenser wired in series as illustrated schematically in figure 46. When a voltage differential is placed across the condenser and then removed, the condenser will attempt to neutralize itself by

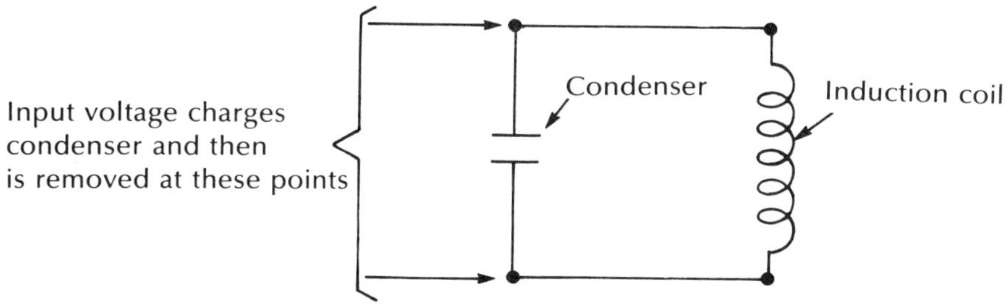

Figure 46
Simple Oscillating Circuit

discharging across the coil. As the current rises from zero to maximum across the coil, a magnetic field is built up. This magnetic field, in the process of building up, acts as a resistance to the current by inducing counter voltages across each loop of this coil. So the buildup of the current and magnetic field is gradual rather than instantaneous, although it still happens within a tiny fraction of a second. As the current reaches its peak and then starts to collapse, the magnetic field also starts to collapse. In this process it induces a supporting voltage, and this then causes the collapse of the current to be streched out rather than instantaneous. The collapsing magnetic field thus forces the current into the opposite plate of the condenser, where it stops when no more magnetic action is present. The condenser is now charged in the opposite direction or polarity from its initial charge. With the condenser now charged again, it will seek to neutralize itself by discharging through the coil again. These oscillations would continue indefinitely if it were not for the normal resistance of the wire and circuitry using up the energy and dissipating it as heat. To sustain the oscillations, a pulse of electricity would have to be given to the circuit to overcome the loss of energy caused by the heat generated. The inductive resistance of the coil does not cause any loss of energy.

As I've described previously, the photonic action is similar to the oscillating circuit with the exception that no condenser or coil exists. But you do have the electrostatic and the magnetic fields that substitute for them. The repulsion of the electrostatic field and the attraction of the magnetic field causes the particles to bounce back and forth as they move through space, resulting in the wave action that they produce.

One would assume that the maximum intensity or amplitude of the oscillations would occur when the magnetic and electric force fields are in balance. Since the particle's maximum electric charge occurs when it is at rest, the velocity at which that charge is reduced to one-half strength would be the point of balance. The magnetic force field would rise to one-half its strength at the same velocity. At the maximum velocity of the particle, the electric charge would be reduced to zero while the magnetic force field would reach its maximum strength.

The current scientific opinion is that the electric charge of a particle remains constant while the mass varies with the velocity. The reason for this is that laboratory experiments have shown that it takes disproportionately more and more power to increase the velocity of an electron. Since they've concluded that the electric charge remains constant, then the obvious assumption is that the mass increases to account for the greater disproportionate power required. However, my opinion is just the opposite. The mass remains constant while the electric charge varies.

If a charged particle generates a magnetic field as it moves, then where does that magnetic force field come from? It can not materialize out of nothing. It must be produced at the expense of the electric charge. If the magnetic field increases with velocity then the electric charge must decrease. At the particle's maximum velocity, the magnetic field must be at its maximum while the electric charge must be reduced to zero. The combined strength of the two forces must equal a fixed value that does not vary. The nature of the force field simply changes from electrostatic to magnetic as the velocity of the particle increases, the mass remaining constant, of course.

Using a formula from Einstein's theory of relativity, the mass of a particle increases with velocity and reaches infinity at the speed of light. Now if we were to use this formula to determine the velocity where the forces balance at one half their maximum strength at the point where the mass is doubled, then the velocity of balance between the electrostatic and magnetic forces would be:

$$m_v = \frac{m_o}{\sqrt{1-(v^2/c^2)}}$$

m_o = mass at rest
m_v = $2m_o$ (rest mass doubled)
v = velocity in question
c = velocity of light

If we rearrange the formula to determine v we have:

$$v = \sqrt{c^2\,[1-(m_o/m_v)^2]} \qquad v = .866$$

This, of course, is assuming that it takes the same amount of power to drive a particle at that velocity whether the mass was doubled or the electric charge was reduced to one half its rest strength, which would create the illusion that the mass was doubled. If the mass remains constant then the electric charge must be reduced to one half. If the electric charge remains constant then the mass must be doubled. Either way, the situation would be the same.

At 86.6 percent of c, we can then assume that the particles oscillations reach their maximum intensity. From that point on it remains at maximum intensity without varying till it reaches the speed of light.

The nature of the oscillations would start to change slightly after the particles exceed the velocity of balance at 86.6 percent of c, but the wave intensity would still remain the same. From that point on, the electric charge continues to decrease from one half strength to zero at c, while the magnetic force field would increase from one half to maximum at c. This means that the particles would be approaching each other closer and closer. Then, at a point still below the speed of light, the particles will start crossing each other's paths when the electric charge has been reduced to such a weakness that it can no longer reverse the direction of the particles, the momentum of the particles overcoming the weak electric charges. But the increasing strength of the magnetic fields will be reversing the direction of the particles now, however, as illustrated in figures 47, 47a, and 47b. At the velocity of light the magnetic force fields will sustain the oscillations at maximum. You'll notice that the particles will interweave at a certain point beyond the velocity of balance of 86.6 percent of c. (Refer to figure 48.)

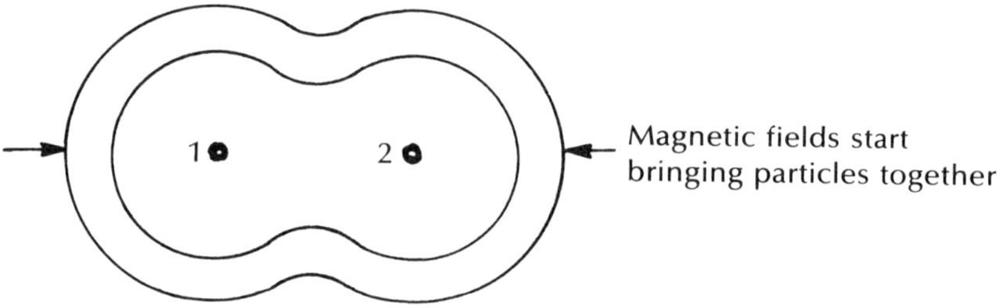

Figure 47
Wave Change beyond Velocity of Balance

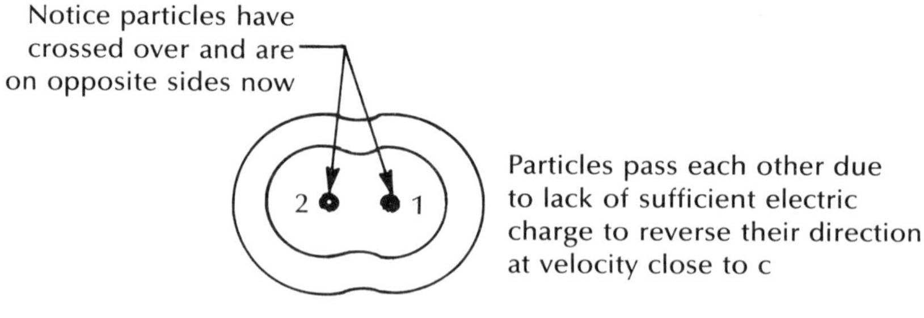

Figure 47a

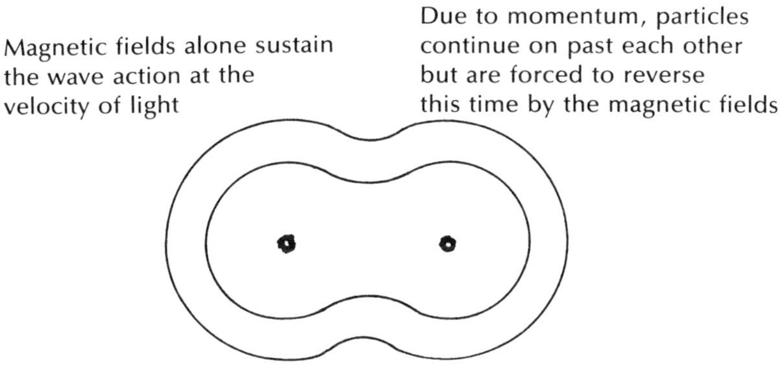

Figure 47b

MAGNETIC FORCE

Magnetic force has to be taken into consideration when evaluating the motion of particles, whether in space or in the atom.

The electrons revolving in the atoms also have a magnetic force field surrounding them. This magnetic field also has some effect on the velocity and size of the orbits, but only when the electron is close to the proton such as in the central regions of the stars, where the atoms are packed together from the enormous gravitational pressure. The revolving electron will cause the proton to spin because of the attraction between the two. The closer the electron is forced to revolve around the proton the faster will be its velocity, and subsequently it will cause the proton to spin faster. The spin of the proton will also generate a magnetic field such as the earth's spin generates. This field will not be as strong as the one surrounding the electron because its spin velocity will be much less than the electrons velocity. However, it still will have some effect between the two particles.

The magnetic fields between the two particles will cause a repulsion because in this instance the proton is in the center rather then revolving around a common axis, as I've explained previously with the positron. At the same time the increased velocity of the electron has caused it to lose some of its electric charge. This loss is replaced by a stronger magnetic field. The result is that the closer an electron approaches a proton, the stronger the magnetic fields generated and the greater the repulsion, which tends to neutralize the electrostatic attraction. This would indicate that these atoms can withstand much higher gravitational pressures than assumed or calculated.

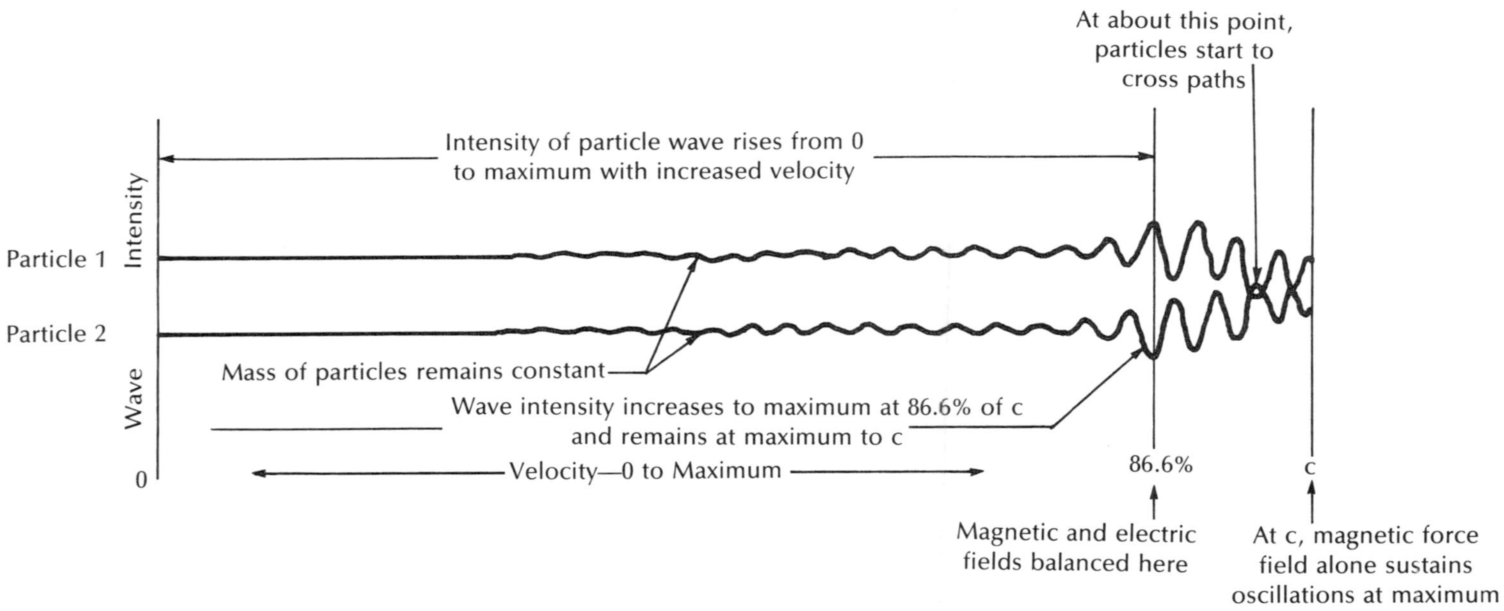

Magnetic field increases with velocity while electric field decreases. At c, magnetic force field is at maximum while electric force field is reduced to zero

Figure 48
Nature of Wave Change of Particle

At the same time, the destruction of a number of electrons by collision would create a certain percentage of free protons within the central cores. The repulsion of the free protons would also create additional central pressure to withstand the gravitational pressure. These free spinning protons could orient themselves so that their magnetic fields could cause some attraction to weaken the repulsion force but that would be very slight. (Refer to figures 49 and 49a.)

The very center of neutron stars would have the highest percentage of free protons. The percentage would gradually reduce to the outer surface of these stars.

With the outer gaseous envelope of the ordinary star burned away after a time, the remaining neutron core would then begin to disintegrate because of the force of repulsion between the free protons, which will be stronger than the force of attraction between the remaining neutrons and protons, and the complete disintegration of the star would follow.

LIFE OF FREE NEUTRONS

Magnetic force may be involved in the disintegration of the free neutrons outside the nucleus of the atom. If we were to assume that the neutron is spinning at high velocity within the nucleus and then

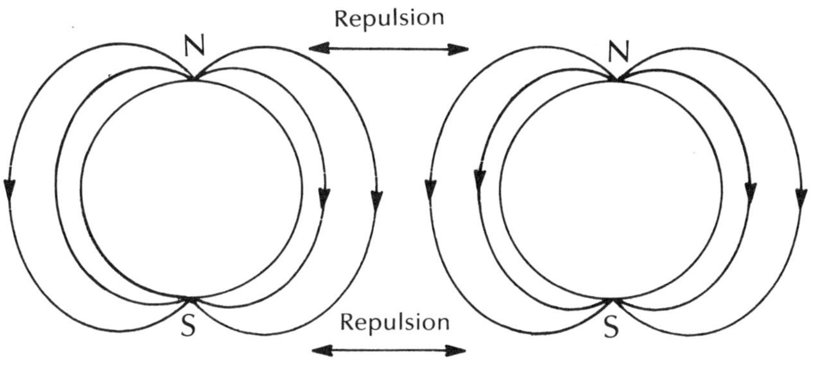

Attraction between north and south poles would cause one proton to flip.

Figure 49
Spinning Protons, Unstable Axis Position

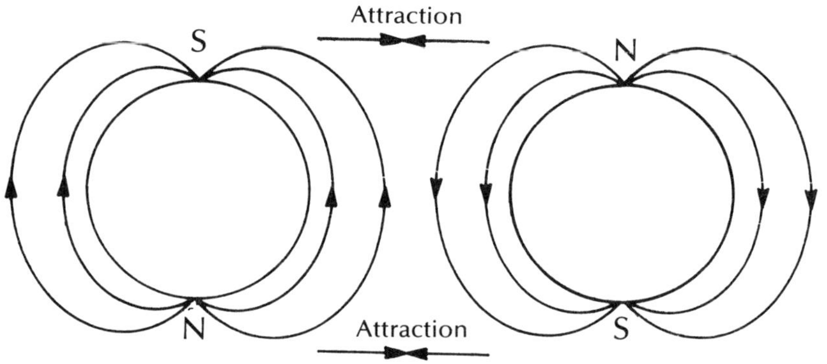

Cancellation of central lines would cause some additional attraction between protons but hardly enough to overpower the electrostatic repulsion.

Figure 49a
Spinning Protons, Stable Axis Position

is separated from it by some means, we would have to take into consideration the magnetic force fields resulting from that spin and the movement of the particle as a major factor in the separation of the neutron into a proton and an electron. First of all, the spin is going to cause a strong repulsion between the proton portion and the electron portion of the neutron, as illustrated in figure 50.

Then the ejection of the neutron and its velocity from the nucleus will also set up some magnetic fields around each component of the neutron and reinforce the repulsion of the magnetic fields from the spin as illustrated in figure 51.

The net result is that the magnetic field may be stronger than the electrostatic and force the two particles to separate.

The direction of the spin's axis relative to the direction of motion away from the nucleus may determine the time it takes for the two particles to separate. With the axis of the spin in line with the direction of motion, the separation would be the fastest. With the axis perpendicular to the line of motion, the separation would be the longest, because of loss of repulsion when the electron is positioned directly on longitudinal line of flight.

The velocity of the spin and the velocity of the flight from the nucleus would have the greatest bearing on the time required for the particles to separate.

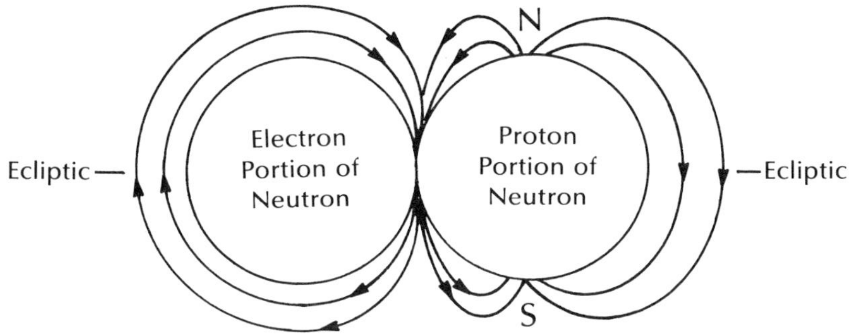

Looking along the plane of the ecliptic

Electron portion attached to and revolving around proton portion of neutron. Electron is revolving but not spinning. Proton is spinning on its axis. Direction of electron is away from observer. Direction of spin would have no bearing on result. Either direction would produce same effect. Notice lines of force between particles moving in same direction. This causes outward pressure between the two particles or repulsion. Centrifugal force should also be considered in the separation of the particles.

Figure 50
Magnetic Fields around Spinning Neutron

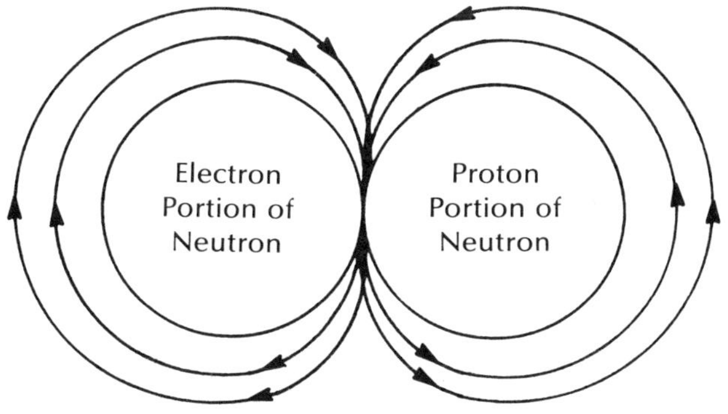

Looking along line of flight away from observer

Flight of particles away from nucleus causes magnetic fields to be set up, which also cause repulsion to reinforce spinning repulsion. Notice how fields between the two particles reinforce each other by moving in the same direction. This causes outward pressure and a repulsion. With the axis of the proton's spin and the electron's rotation in line with the flight of the particles, the magnetic forces would be the greatest and the separation of the particles the quickest. With the axis of the spin perpendicular to the line of flight, the separation of the particles would take the longest.

Figure 51
Magnetic Fields around Neutron in Flight

Addendum

This addendum is devoted to the clarification of certain terms used in astronomy. Only those which I think require clarity of definition are mentioned.

SPACE

I think this term is most confusing because of the various definitions attributed to it, such as curvature of space, hyperbolic space, spheroidal space, flat space, one-, two-, three-, and four-dimensional space. They are confusing because they do not represent the true meaning of space.

Space is simply the void or nothingness between the particles and forms of matter. Space has no form; only matter has form. Space does include the small quantity of matter that occupies it. In other words, matter is also considered a small part of space although practically all space is a void.

There is only three-dimensional space. It is infinite in its dimensions, which means there is no limit as to how far it extends. Only matter has limitations because it has form. Our universe could be limited in dimension. That would be when the matter in our universe ceases to extend beyond a certain boundary. However, space would still continue to infinity.

One-, two- and four-dimensional space are really nonexistant. They only represent concepts or ideas. One-dimensional space would be represented by a straight line because it has but one dimension. Two-dimensional space would be represented by a flat square or rectangle because it would have just two dimensions.

Four-dimensional space is inconceivable because, literally, no fourth dimension really exists. Time is substituted for the fourth dimension but time is not really a dimension. Time is explained more thoroughly later on. By using a three-dimensional space such as a cube, and selecting some point in that cube to represent a period of time, you then have to pinpoint it in that cube. The location of that point has to be represented by three dimensions or coordinates which are height, width, and length. To represent another point as an earlier or later moment in time nearby, that point also has to be pinpointed by three dimensions or coordinates. The period only represents an earlier or later time period, but each period has to be pinpointed by three-dimensional coordinates. So the fourth dimension as a measure of space is really a nonexistant entity.

Curvature of space is also nonexistent. The curvature of space is really a substitution for the curvature of the motion of a body of matter as it moves through the various gravitational fields in the universe. In the case of the small particles, it would be their curvature of movement through the electrostatic or magnetic force fields within the atom or through space. Curvature of space then represents the movement of bodies through space.

Hyperbolic space is represented by a saddle-shaped curvature. The real nature of a hyperbola relative to space is a curve formed by a body moving through a gravitational field at a velocity sufficiently high enough for it to avoid being captured and held in orbit. Bodies captured by gravitational fields move in elliptical and closed orbits. The hyperbola, of course, would not be a closed orbit. A geometric example of a hyperbola is given in figure 54. Figures 52 and 53 show an ellipse and a parabola, respectively.

Spheroidal space is the space taken up by a large body of matter such as a planet or star and really pertains to matter rather than space, although matter does occupy space. All these large bodies may be slightly elliptical due to their spin, but it isn't visible to the naked eye.

Flat space is really two-dimensional and can pertain to the movement of bodies in orbit. Basically, two-dimensional space exists only on paper because the real movement of bodies through space would be in a helical or spiral form, when the totality of their movement is taken into account. This would be three-dimensional.

MATTER, ENERGY, AND FORCE

Energy is simply matter in motion. You cannot have energy without matter. However, matter can theoretically exist without energy when in an immobile state. In space, however, that is an impossibility also because matter is always in a state of motion as a result of the various forces that exist. These forces, of course, are the gravitational, electrical, and magnetic. These forces also cannot exist without matter because they are an inherent property of matter. The earth's movement around the sun is energy brought about by the gravitational force

Conic Sections Identifiable with Closed Orbits

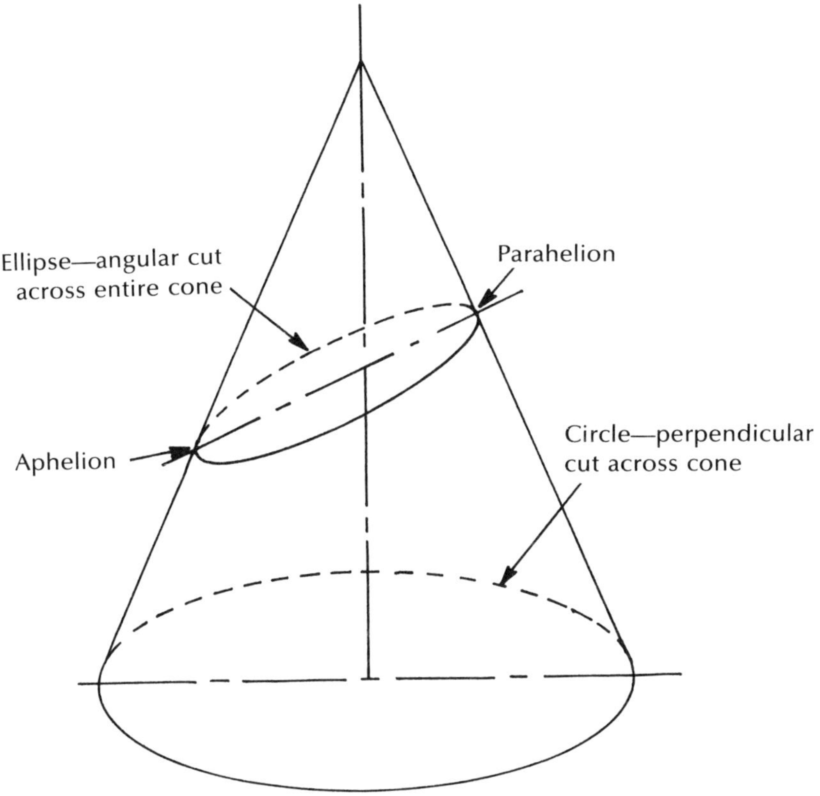

A body with an elliptical orbit has a variable motion. The more acute the angle, the greater the eccentricity and the greater the variability of the motion. A body with a circular orbit, of course, would have uniform motion throughout. However, this type of orbit would be very unlikely.

Figure 52
Circular and Elliptical Orbits

between them. The electron's movement around the proton is also energy resulting from the electrostatic force between them.

Energy cannot be converted into matter or matter into energy. I know that contradicts current scientific opinion, but matter and energy are two different things. When matter is put into motion it is in an energy state. Pure energy without matter cannot exist. That is why I would like to reemphasize here that the photon is not a packet of energy but a tiny particle of matter.

Latent energy is contained in each particle of matter by the cohesive forces within that particle. Once that cohesive force is broken by the collision of the particle, then the tremendous energy released is the result of the repulsion between the electrostatic forces within each individual particle. Of course, the energy then is in the motion of these fragmented particles.

Conic Section Identifiable with the Parabola.

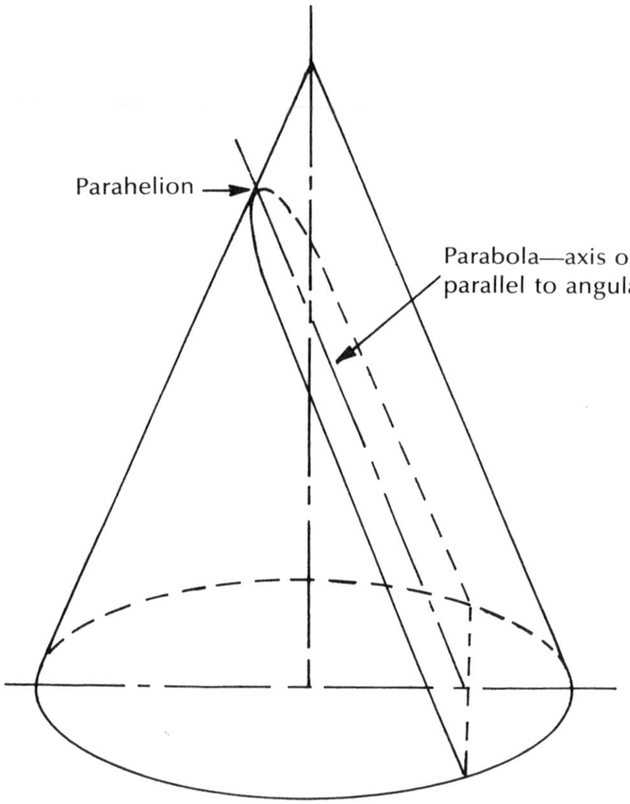

Curvature of motion of this body would be at the threshold velocity of escape. The motion, of course, would be variable and the orbit would be open.

Figure 53
Parabola

Conic Section Identifiable with the Hyperbola.

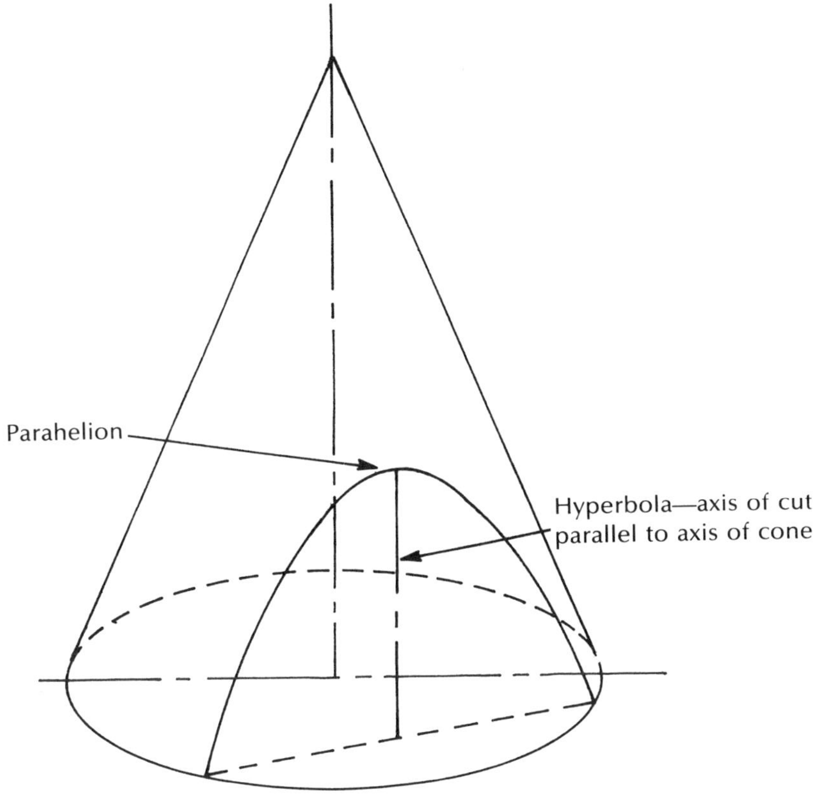

Curvature of motion of a body forming a hyperbola would have a velocity beyond escape. The orbit would be open and the motion would be variable.

Figure 54
Hyperbola

TIME

Time is simply uniform motion. Just as distance is based on a uniform unit of length, time is based on a uniform unit of unwavering motion. The length of our day is based on the spin of our earth on its axis, and it denotes one complete revolution. Because of the earth's size, there is very little variation to its duration of spin, if any does exist. The length of our year is based on the movement of our earth in its orbit around the sun. Again, there is very little variation to the velocity of its movement in that orbit. There is a slight variation of the motion in different parts of the orbit due to its eccentricity, but not to the length of the year.

That slight variation in motion has nothing to do with the length of the months. The length of the months are more or less based on the revolution of the moon around our earth. The length of the months do not coincide exactly with the moon's revolution because there is a disparity with the length of the year. The moon completes 12.38 revolutions during each year. Because of that, the phases of the moon do not fall on the same days of each month as a result.

A light year, although it sounds like a unit of time, is actually a measure of distance. It is the distance that light travels in one year, and the term is used for the sake of convenience because of the enormous distances in astronomical measurements.

Bibliography

ASTRONOMY

Arp, Halton C. "Related Galaxies with Different Red Shifts." *Sky and Telescope Magazine,* April 1983.

Kaufmann, William J. III. *Galaxies and Quasars.* San Francisco: W. H. Freeman and Co., 1979.

Pasachoff, Jay M. *Contemporary Astronomy.* Philadelphia: W. B. Saunders Co., 1977.

Santage, Allan. *Hubble Atlas of Galaxies.* Washington, D.C.: Carnegie Institute, 1961.

The Editors of *Scientific American. Cosmology + 1.* San Francisco: W. H. Freeman and Co., 1977.

AURORA BOREALIS AND VAN ALLEN BELTS

Akasofu, S. I. "Aurora Borealis—Amazing Northern Lights." *Journal of the Alaska Geographic Society, vol.* 6, no. 2 (1979).

International Union of Geodesy and Geophysics. *International Auroral Atlas.* Edinburgh: Edinburgh University Press, 1963.

Van Allen, James A. *Origins of Magnetospheric Physics.* Washington, D.C.: Smithsonian Institution Press, 1983.

PHYSICS

Born, Max. *The Restless Universe.* New York: Dover Publications, 1951.

Hughes, Donald J. *Neutron Story.* Garden City, N.Y.: Doubleday and Co., 1959.

Young, Louise B. *Mystery of Matter.* New York: Oxford University Press, 1965.

Index

Absolute magnitude, 27
Absorption lines, 65
Age of universe, 77, 79
Alpha particles, 3
Andromeda galaxy, 66, 69–71
Aphelion, 12
Apparent magnitude, 27
Aristotle, 76
Asteroids, 10, 12
Atmosphere, 34
Atom, 91, 92
Atomic energy, 5
Aurora borealis, 11, 34

Background radiation, 72
Beta particle, 3
Big bang, 44, 63–83
Binary star, 27
Black hole, 84–87
Blue shift, 66, 69–71

Calcium lines, 66
Celsius, 72
Centaurus A, 32
Cepheid variable, 26–29
Cohesive bond, 5
Comets, 10–12
Continuous spectrum, 65
Copernicus, Nicolaus, 76
Cosmological principle, 63
Crab Nebula, 18–26
Curvature of space, 110
Cygnus A, 29–33

de Broglie formula, 6
Delta of Cepheus, 27
Density, 75
  of particles, 87–88

Dumbbell Nebula,18

E layer, 34, 35, 36
Earth, 4
Einstein mass formula, 101
Einstein X-ray satellite, 24
Electro-static force, 4, 5, 92
Electron(s), 3, 4, 5, 63, 77, 91, 103
  density, 88
  mass, 87
  size, 88
Ellipse, 110
Emission lines, 65
Energy, 110
Event horizon, 84

F layer, 34, 35, 36
Force(s), 110
  electro-static, 4, 5, 92
  magnetic, 10, 93

Galaxies, 45
  Andromeda, 66, 69–71
  barred spiral, 50, 52, 58–61
  classification of, 50–52
  clusters, 53, 69
  desintigration of, 46
  elliptical, 50, 53
  formation of, 52–58
  irregular, 50, 52
  Local Group, 69
  Milky Way, 46, 52, 63, 69–71
  spiral, 50, 52, 53, 58
Gamma rays, 3
Gobular clusters, 52

H & K lines of calcium, 66
Hubble constant, 75, 79

Inverse square law, 71
Isotropic universe, 74

Jupiter, 4, 10, 12, 14

Kelvin, 72

Light year, 114
Local Group, 69

M1, 18–24
M27, 18
M31, 66, 69–71
M33, 58
M57, 14
M87, 32, 53
M97, 18
M104, 55
Magnetic field, 93, 96, 99–101
Magnetic force field, 10, 93, 97, 101, 103–105
Mass of
  electron, 87
  neutron star, 1
  photon, 6
  proton, 87
  sun, 3
  white dwarf, 1
Matter, 91, 109
  regeneration of, 45
Meteoroids, 10–12
Meteors, 34
Michelson and Morley experiment, 6–9
Milky Way galaxy, 46, 52, 63, 69–71

Nebula,
  Crab, 18–24
  Dumbell, 18
  Eskimo, 18
  Owl, 18
  Ring, 14
Negative ions, 47, 48
Neptune, 10
Neutron, life of, 4, 105
Neutron star, 1, 4, 23, 24, 75, 84, 86, 89
NGC,
  598, 58
  1199, 68
  2392, 18
  4565, 58
  4594, 55
  5128, 55
  7293, 16
  7603, 66
  7741, 61
Northern lights, 34
Nova, 14

OH 471, 81
Olbers paradox, 71–72
Orbits of bodies, 110
Oscillating circuit, 99
Owl Nebula, 18

Parabola, 110
Parsec, 27
Parahelion, 12
Photon(s), 2, 96–103
  mass of, 6
  velocity of, 3
Planck's constant, 6
Positron, 99
Primordial background radiation, 72–74
Proton(s), 2, 5, 6, 45, 77, 91
  density, 87
  mass, 87
  size, 87
Pulsar, 20–23, 24

Quasars, 68, 79, 81

Radio astronomy, 20
Red shift, 64–69
Right-hand rule, 93
Ring Nebula, 14

Saturn, 4, 10
Singularity, 63, 84
Sirius, 11
Solar flares, 10, 12

Solar system compared to atom, 92
Sombrero galaxy, 55
Sound wave, 3, 64
Space, definition of, 109
Spectrograph, 66
Spectrum, 64, 66
Speed of light, 77
Spiral galaxy, 50, 52, 53, 58
Star(s),
  buildup, 47, 48, 49
  Cepheid variables, 26–29
  desintigration of, 45
  heat distribution, 87
  neutron, 1, 4, 23, 24, 75, 84, 86, 89
  types of, 4
  white dwarf, 1, 3
Steady state theory, 45, 64
Strobiscopic device, 21
Sun(s), 1–4
  mass, 3, 5
  mass loss, 3, 5
  temperature, 2
Sunspot cycle, 11
Sunspots, 10
Supernova, 14

Temperature, 73
  3C9, 3C48, 3C123, 3C147, 81
Time, definition of, 113
Triangulum galaxy, 58

Universe,
  age of, 79
  big bang, 44, 63–83
  definition of, 44
  perpetual, 44–61

Van Allen radiation belts, 36
Variable stars, 26
Velocity of light, 75

Wave nature of particles, 91–105
White dwarf, 1, 3
Weight of particles, 87, 88

X ray, 46
  satellite, 24